Starlight, Starbright: Are Stars Conscious?

Starlight, Starbright: Are Stars Conscious?

GREG MATLOFF AND C BANGS

Greg Matloff
Jun 23, 2017

Dr. Gregory L. Matloff
Professor
Physics Department, New York City College of Technology, CUNY

ISBN: 978-0-9934002-1-6

First published 2015 by Curtis Press

Cover design: Jim Wilkie

Visit Curtis Press at *www.curtis-press.com*

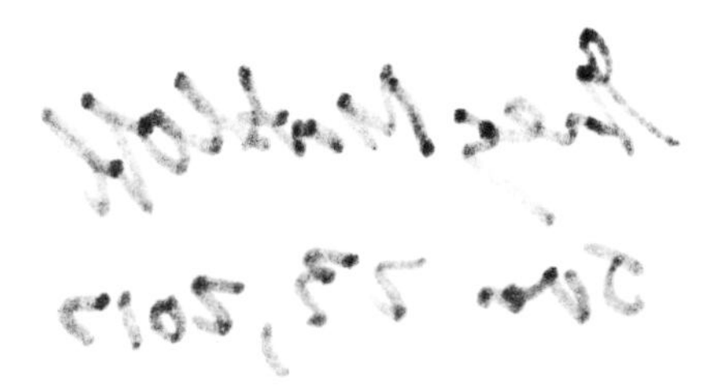

Contents

Dedication

This volume is dedicated to the memory of three extraordinary people. Each in his own way was instrumental in providing a framework for the ideas explored here. First, here's to Olaf Stapledon! In *Star Maker* and other novels, he brought the concept of stellar consciousness to a wide audience. Next, we remember Arthur C. Clarke for his contributions in making the Universe in all its vastness and wonder accessible to the entire world. Finally, we will never forget Evan Harris Walker for his pioneering studies in the physics of consciousness, his friendship and support.

Preface

There are many published books dealing with the phenomenon of consciousness. Indeed, it may be the most thought about and discussed topic of all. Although our own consciousness is perhaps the only thing we can be absolutely certain of, it is paradoxically a very difficult concept to define.

I have nothing to add to the age-old efforts to explain consciousness. But in this book I explore the possibility that whatever it is, it may be present in other entities as well as in organic life—namely, the stars.

Other books have been published over the centuries to explore the possibility of stellar consciousness. But many of these have been authored from the point of view of Revealed Knowledge. Others have approached the concept of conscious stars from the direction of Deductive Knowledge. Although this book refers to stellar consciousness in myth, poetry and literature, it is primarily concerned with application of the scientific approach.

The reader will note that this book is divided into four parts. Those interested in ancient myths and traditions related to the concept of stellar consciousness may be most interested in "Part I: Ancient nights." Those who follow the progress of quantum consciousness theory may concentrate on "Part II: Of neurons, tubules, and molecules" and the final chapter. Readers with a literary interest might concentrate on "Part III: Fictional bright stars" and the final chapter. Those interested in observational evidence might start with "Part IV: The astronomer's search."

As the reader may recall, the source of Revealed Knowledge is either direct contact with some deity or the human unconscious. Deductive knowledge is the result of rigorous logic. Although deduction is necessary in science for the construction of hypotheses, experimental or observational inductive tests are required before these constructs can emerge as successful theories. No claim is made here that all knowledge must be inductively testable in scientific terms. But since I am a scientist, I prefer to approach phenomena from a scientific viewpoint.

I was led to propose conscious stars in 2011 as a scientific hypothesis in response to the ongoing crisis of dark matter in astrophysics, as is discussed in the "Introduction." Gravity theory seems to require a large amount of unseen matter to explain the motions of very distant stars and galaxies. But all attempts to locate this material have failed.

Physical science may be at a similar turning point to that of 1900, when observations of the speed of light in vacuum were beginning to upset the established paradigms of classical physics. This unsettling situation began to be resolved when Albert Einstein proposed the special theory of relativity in 1905.

Stellar consciousness, as defined here, is very simple. To partially explain the kinematical anomalies discussed in this book, a star must possess a simple herding instinct. Perhaps its consciousness is no more advanced than that of a slime mold amoeba.

In 2011, I thought that it might be necessary to revive and resolve the debate on psychokinesis (PK) to explain how a conscious star might alter its trajectory. PK is the alleged ability of certain individuals to move objects at a distance by the power of will. But observations of unidirectional, high-velocity jets of material ejected from some young stars provide an alternative explanation. But since I have been privileged to know capable people on both sides of the PK debate, this subject is discussed in its possible relation to anomalous stellar motions.

As well as publishing the original conjecture in a scientific journal (*Journal of the British Interplanetary Society—JBIS*), I submitted it to an astronomical blog (*Centauri Dreams*). Many of the responses to the blog entry provided creative approaches to testing both stellar consciousness and PK. Happily (for me), a suggested alternative to stellar consciousnees has not passed its first observational test.

The jury is still out and a verdict has not been returned. But a new European astronomical spacecraft, *Gaia*, has been launched and is now on station. Within a few years, intimate knowledge of the motions of about a billion stars in our galaxy may be available from this spacecraft. I await these results eagerly.

Foreword

As an astrophysicist, I appreciate the value of revolutionary ideas in science, but I also understand the unique challenges that confront efforts to advance them against the resistance of established scientific paradigms. It is important to point out that all established scientific paradigms are founded on the assumptions that were invoked by scientists to construct their hypotheses and theories. And it is the assumptions that are the most prone to replacement because new experimental evidence or improved theoretical developments come along which show the original assumptions to be wrong or incomplete. While my coworkers and I are exploring the overlap between the multi-layered structure of quantum vacuum zero-point fluctuations, quantum entanglement and teleportation, the sub-quantum information domain, and the physics of consciousness, the creators of this book are going far beyond this by investigating the possibility of stellar consciousness.

Stars might be conscious? That sounds patently absurd. But this is only absurd at face value given the long-cherished scientific *assumption* that only living beings are conscious and not nonliving matter, much less a massive astronomical-sized, hot, dense ball of plasma that is undergoing nuclear fusion. How can stars be conscious if they are not living beings? That is precisely the question that this book addresses, and the authors take the risk of seriously considering the evolving astronomical evidence for anomalous kinetic stellar motions in our galaxy that could turn out to be volitional. If real, then such stars might be using psychokinesis to project unidirectional jets of plasma in order to produce thrust. We do not yet have a physics theory of consciousness so the long-held paradigm that nonliving matter cannot possibly be conscious could be wrong.

The suggestion that stellar matter could be conscious is not really a stretch too far for science. Dmitri Krioukov and coworkers at the University of California-San Diego and the University of Barcelona suggest that the universe may be growing in the same way as a giant brain with the electrical firing between brain cells being "mirrored" by the shape and structure of galactic matter throughout the expanding universe (D. Krioukov et al., *NATURE Scientific Reports*, **2**: 793, 2012). The results of their theoretical and computer model simulation suggests that natural growth dynamics—the way that systems evolve—are the same for different kinds of networks whether it is the social networks on the internet, the human brain, or the universe as a whole. The discovered equivalence between the growth of the universe and complex networks strongly suggests that unexpectedly similar laws govern the dynamics of these very different complex systems. The observed structural and dynamical similarities among different real networks suggest that some universal laws might be in action, although the nature and common origin of such laws remain elusive. A physics of consciousness could be the key to fully understanding this problem.

On the other hand, quantum mechanics predicts that anything that has the slightest chance of existing is virtually certain to exist, so that given enough time any vanishingly improbable event will occur. Taken to its logical conclusion, there is an extremely tiny probability of self-aware, disembodied consciousnesses, or "brains," that spontaneously emerge from the quantum vacuum zero-point fluctuations of "empty" space everywhere across the universe (A. Albrecht and L. Sorbo, *Phys. Rev. D*, **70**: 063528, 2004; J. Garriga et al., *J. Cosmology and Astroparticle Phys.*, **0601**: 017, 2006; D. Page, *Phys. Rev. D*, **78**: 063536, 2008; A. De Simone et al., *Phys. Rev. D*, **82**: 063520, 2010). These are called "Boltzmann brains," and over the entire history of the universe it is possible that such brains would outnumber consciousnesses such as ours. Boltzmann brains are named for the physicist Ludwig E. Boltzmann (1844–1906) who hypothesized that the universe is observed to be in a highly improbable non-equilibrium state because only when such states randomly occur can brains exist to be aware of the universe. So quantum mechanics, taken together with non-equilibrium thermodynamics and modern cosmology, appears to already provide a physics theory of consciousness. It could be possible that Boltzmann brains inhabit volitional stars and possibly even the entire universe itself, and it may even be possible that Boltzmann brains give consciousness to all sentient life forms.

Boltzmann brains could also be the source of a very unusual, but historically well-known and well-documented phenomenon called psychokinesis (PK). PK has been recorded in very ancient human history, but it is among the many types of human potentials that were explored during experimental research in the field of enhancing human performance for a military program called JEDI, which ran in the 1980s under the auspices of the U.S. Army Intelligence and Security Command and the Organizational Effectiveness School, and was sponsored by a U.S. government interagency task force (J. B. Alexander, R. Groller, and J. Morris, *The Warrior's Edge*, William Morrow & Co., New York, 1990; E. W. Davis, "Teleportation Physics Study," *AFRL-PR-ED-TR-2003-0034*: 55–62, 2004). As strange as Boltzmann brains may sound to scientists and non-scientists alike, PK is an even stranger phenomenon. PK generally designates the movement of objects through other physical objects or over great distances. Telekinesis (TK) is a form of PK which describes the movement of stationary objects without the use of any known physical force. And PK/TK is essentially the direct influence of mind on matter without any known intermediate physical energy, physical contact, or instrumentation. Rigorously controlled scientific laboratory studies of PK/TK have been performed and/or documented by L. E. Rhine (1970), H. Schmidt (1974, 1987), H. E. Puthoff and R. Targ (1974, 1975), J. B. Hasted et al. (1975), R. Targ and H. E. Puthoff (1977), C. B. Nash (1978), S. Shigemi et al. (1978), J. B. Hasted (1979), G. B. "Jack" Houck (1984; *http://www.jackhouck.com/*), B. B. Wolman et al. (1986), W. Giroldini (1991), L. R. Gissurarson (1992), D. Radin (1997), C. T. Tart et al. (2002), R. Shoup (2002), and J. B. Alexander (2003) (the full references are listed in my U.S. Air Force report, "Teleportation Physics Study," *AFRL-PR-ED-TR-2003-0034*: 63–72,

2004). My Air Force report also describes the laboratory PK research that was supported by the Soviet Union, the Warsaw Pact, and the People's Republic of China military and national science academy institutes. These studies demonstrated that PK/TK phenomenon cannot be conjured by magicians in staged magic show acts. If stars are truly conscious, then they might be using PK to project a single jet of plasma out into space in order to produce thrust to move beyond their confined Keplerian orbital motion in the galaxy.

My own theoretical research has led me to conclude that underlying the universe is a sub-quantum (SQ) information domain whereby the vacuum of spacetime is teeming with information as well as the ubiquitous complex multi-layered quantum vacuum zero-point fluctuations. And SQ information is what determines what manifests physically. SQ information might be Fisher Information (B. R. Frieden, *Physics from Fisher Information: A Unification*, Cambridge Univ. Press, Cambridge, U.K., 1999; *Science from Fisher Information: A Unification*, 2nd ed., Cambridge Univ. Press, Cambridge, U.K., 2004) or it might be quantum holographic universe information (J. D. Bekenstein, *Sci. Am.*, **289**: 58–65, August 2003). The jury is still out on which model best describes physical reality and the universe. But Fisher Information is the only theoretical framework so far that provides an approach to understanding consciousness. Like quantum entanglement, SQ information is nonlocal such that information and images can be robustly teleported via instantaneous quantum connections across spacetime. Nonlocal entanglement is common in nature and is not just a low-temperature or microscopic laboratory phenomenon—the quantum domain is not limited to microscopic size and mass scales or to brief instances of time. Several years of recent experimental research has demonstrated that quantum entanglement and teleportation, along with other quantum effects, have been observed in a growing number of large macroscopic systems as well as ones contained in high-temperature and high-pressure environments, including in living organisms (V. Vedral, *Sci. Am.*, **304**: 38–43, June 2011). Furthermore, astronomers have observed and catalogued macroscopic quantum matter of astronomical size, temperature, pressure, and mass such as neutron stars, quark stars, and white dwarf stars which are composed of Fermi-degenerate gases of nucleons or quarks and electrons. The combined experimental and theoretical research concludes that quantum entanglement gives spacetime its structure because causal order is not a fundamental property of nature.

My own research is concerned with the question of how might biological systems exploit SQ information and whether it is integral to consciousness. I also want to know what sort of information structures might exist in SQ information space. Could they be qualia (thought forms), Boltzmann brains, or the complex multi-layered quantum vacuum zero-point fluctuations? It turns out that quantum vacuum zero-point fluctuations are correlated (i.e., quantum entangled). Therefore, the SQ information space might be an SQ entanglement field, and its information structures might be the quantum vacuum fluctuations which can produce Boltzmann brains as previously discussed. Spatiotemporal SQ information structures should be required to interact with matter because matter is spatiotemporal. The synchronizing of spatiotemporal joint probability distributions in matter could cause many anomalies that involve no transfer of energy such as what is observed in, for example, synchronicities in nature, PK, remote perception, thought projection, and telepathy.

Thomas Kuhn wrote, "Discovery commences with the awareness of anomaly, that is, with the recognition that nature has somehow violated the pre-induced expectations that govern normal science" (T. S. Kuhn, *The Structure of Scientific Revolutions*, 2nd ed., Univ. of Chicago Press, Chicago, 1970, pp. 52–53). He described scientific discovery as a complex process in which an observed anomalous fact of nature is accepted and then followed by a change in scientific paradigm that makes the new fact no longer an anomaly. Science is a conservative activity. Such conservatism has the short-term asset of allowing each current paradigm to be articulated so clearly that it can serve as an organizing principle for the multitude of effects that scientists observe, but it can sometimes prevent the adequate recognition of observations that do not fit within the current theory. In addition, scientists may be reluctant to accept new theories, or recognize anomalies in the old theories, for the purely psychological reasons that the familiar is often more comfortable than the unfamiliar, and that inconsistencies in belief are uncomfortable—which is called cognitive dissonance. If unexplained facts can be glossed over or reduced in importance or simply accepted as givens, the possible inadequacy of the current theory does not have to be confronted. Then, when a new theory gives a compelling explanation of the previously unexplained facts, it is safe to recognize them for what they are. Such "retrorecognition" phenomenon of certain scientific anomalies has happened again and again throughout the history of science (A. Lightman and O. Gingerich, *Science*, **255**: 690–695, 1991). Stellar consciousness and PK most certainly fall into the category of retrorecognition phenomenon.

In this book, Greg Matloff and C Bangs present preliminary evidence for a new anomaly in astronomy that potentially could become the next retrorecognition phenomenon that upends a few established scientific paradigms. To build their case for stellar consciousness and explore what consciousness is, they take the reader through a grand tour of these concepts that includes the interweaving viewpoints of the shaman, the mystic, the astrologer, the philosopher, the poet, the novelist, the psychic, the neuroscientist, the molecular biologist, the physicist, the mathematician, and the astronomer. This tapestry of exploration is woven together by the rich artwork by C Bangs who expresses through her art the ideas and thoughts that are inexpressible by words. It is hoped that this book will inspire students and professionals, as well as the interested layperson, to explore this unusual new stellar phenomenon and initiate their own studies into consciousness, spacetime, cosmology, and their interrelationships.

Eric W. Davis, Ph.D., FBIS, AFAIAA
Institute for Advanced Studies at Austin, Austin, TX
August 2015

Acknowledgments

We are especially grateful to the British Interplanetary Society for having us present early versions of the work described here on two separate occasions and for including a peer-reviewed paper by author Greg Matloff in an issue of *Journal of the British Interplanetary Society (JBIS)*. Also, thanks to Paul Gilster for including a longer version of Greg's paper in his *Centauri Dreams* blog and to the many people who commented on the blog entry. We also appreciate our publication on this topic in the *Baen Press* online science magazine.

Our presentations on stellar consciousness combining art and science at a meeting of the LASER art/science salon in New York City and at the Liberty Con and Dragon Con science fiction conventions were fun as well, thanks to the hard work of the organizers.

Much of the original art by C Bangs that has been integrated into the chapter frontispieces is on display at CENTRAL BOOKING art space, at 21 Ludlow Street in Manhattan. C appreciates the encouragement of gallery director and curator Maddy Rosenberg.

The library staff of New York City College of Technology, in Brooklyn, NY is thanked for securing an inter-library loan of Erich Jantsch's book, *The Self-Organizing Universe*. We also thank the research library staff of the National Museum of the American Indian in New York City for their assistance in locating books on Native American science and archeoastronomy.

Finally, we thank publisher Neil Shuttlewood, founder of Curtis Press, for his encouragement and assistance and the staff of that organization for producing the final product.

March 2015 *Greg Matloff and C Bangs*

About the author and artist

Greg Matloff is a leading expert in possibilities for interstellar propulsion, especially near-Sun solar sail trajectories that might ultimately enable interstellar travel. He is also a professor with the Physics Department of New York City College of Technology, CUNY, a consultant with NASA Marshall Space Flight Center, a Hayden Associate of the American Museum of Natural History, and a Member of the International Academy of Astronautics. He co-authored with Les Johnson of NASA and C Bangs *Paradise Regained* (2009), *Living Off the Land in Space* (2007), and has authored *Deep-Space Probes* (Edition 1: 2000 and Edition 2: 2005). As well as authoring *More Telescope Power* (2002), *Telescope Power* (1993), *The Urban Astronomer* (1991), he co-authored with Eugene Mallove *The Starflight Handbook* (1989). His papers on interstellar travel, the search for extraterrestrial artifacts, and methods of protecting Earth from asteroid impacts have been published in *Journal of the British Interplanetary Society*, *Acta Astronautica*, *Spaceflight*, *Space Technology*, *Journal of Astronautical Sciences*, and *Mercury*. His popular articles have appeared in many publications, including *Analog* and *IEEE Spectrum*. In 1998, he won a $5000 prize in the international essay contest on ETI sponsored by the National Institute for Discovery Science. He served on a November 2007 panel organized by Seed magazine to brief Congressional staff on the possibilities of a sustainable, meaningful space program.

Greg Matloff is a Fellow of the British Interplanetary Society, has chaired many technical sessions, and is listed in numerous volumes of *Who's Who*. In 2008 he was honored as Scholar on Campus at New York City College of Technology. His most recent book, co-authored with Italian researcher Giovanni Vulpetti and Les Johnson, is *Solar Sails: A Novel Approach to Interplanetary Travel*, Springer (2008, 2015). In addition to his interstellar travel research, he has contributed to SETI (the Search for Extraterrestrial Intelligence), modeling studies of human effects on Earth's atmosphere, interplanetary exploration concept analysis, alternative energy, in-space navigation, and the search for extrasolar planets. His website is *www.gregmatloff.com*

C Bangs' art investigates frontier science combined with symbolist figuration from an ecological feminist point of view. Her work is included in public and private collections as well as in books and journals. Public collections include the Library of Congress, NASA's Marshall Spaceflight Center, British Interplanetary Society, New York City College of Technology, Pratt Institute, Cornell University, and Pace University. The "I Am the Cosmos" exhibition at the New Jersey State Museum in Trenton included her work, *Raw Materials from Space* and the *Orbital Steam Locomotive*. Her art has been included in eight books, two peer-reviewed journal articles, several magazine articles, and art catalogs. Merging art and science, she worked for three summers as a NASA Faculty Fellow; under a NASA grant she investigated holographic interstellar probe message plaques. Her recent book collaboration with Greg Matloff, *Biosphere Extension: Solar System Resources for the Earth* was recently collected by the Brooklyn Museum for their artist book collection. *Harvesting Space for a Greener Earth* published by Springer came out at the end of April 2014. She is represented at the New York City Gallery CENTRAL BOOKING Art Space.

Since 1995 she has included quantum equations and diagrams by quantum consciousness physicist Evan Harris Walker in her paintings, after making his acquaintance in 1991. They exchanged ideas on the reality or nonreality of space-time and on his innovative theories concerning the relationship between quantum mechanics and consciousness that lasted until his death in 2006. These equations function as design elements and refer to the interconnectivity of everything and the relationship of time to space. Her website is *www.cbangs.com*

Ad Astra C Bangs

INTRODUCTION

A lesson for the professor

Hot sun, cool fire, tempered with sweet air,
Black shade, fair nurse, shadow my white hair:
Shine, sun; burn, fire: Breathe, air, and ease me:
Black shade, fair nurse, shroud me, and please me:
Shadow, my sweet nurse. Keep me from burning"

George Peele, *Bethsabe Bathing*

Sometimes, college professors can learn a great deal from their students. Such was the case for me a few years ago. What I then learned from a student is one of the roots of this book.

As part of my responsibilities as a tenured professor in the Physics Department of New York City College of Technology (which is located in downtown Brooklyn and is part of the City University of New York), I had developed an astronomy program. Tailored to fit the requirements of liberal arts majors, this freshman/sophomore level course is organized in two consecutive semesters. As is customary in such courses, the first semester is devoted to astronomical history, physical science theory, and details of our Solar System. The second semester includes consideration of the stars, galaxies, cosmology, extrasolar planets, life in the Universe, and other topics in astrophysics.

A DARK DISCUSSION

In an Astronomy 2 lecture late in the semester, I was presenting the current state of the hunt for dark matter. This hypothetical substance, which theoretically accounts for about three quarters of the mass in the Universe, has been invoked to explain strange aspects of the motions of distant stars and galaxies. But it has proven to be extremely elusive. Moreover, attempts to modify gravity theory to accommodate this phenomenon have not succeeded.

A student raised his hand. Since I encourage class participation, I recognized him. "Let's see," he exclaimed. "The stars are observed to move around the galaxy's center as if they are attached to the spokes of a wheel, with those farther from the center moving faster." I agreed and suggested that he continue.

"So it is assumed to explain this phenomenon that an invisible cloud of matter surrounds the visible component of the universe. The particles that make up this cloud have not been detected although their combined mass greatly exceeds that of the Universe's visible mass. But all attempts to locate this material has failed and spacecraft trajectories seem unaffected by its presence in the Solar System." Once again, I confirmed that he was correct.

"Furthermore, attempts to modify Newton's and Einstein's gravity equations to account for this effect have failed." I told him that he was on a roll.

The class waited expectantly for his conclusion, as did I. "I think that we have to make an attempt to save the system, kind of like what went on around 1900 when the anomalies that led to relativity theory were recognized. Perhaps the whole attempt to postulate and search for new particles that solve the dark matter problem will fail. Instead, physics may eventually have to alter its organizing paradigm as happened back then."

Two angels rested on my shoulders as he presented his case for dismissing the dark matter hypothesis with radical surgery. My dark angel suggested that I shut this student up. He was deviating from the lesson plan and taking up valuable class time. After all, we had only a few more lessons to complete the syllabus.

But my bright angel offered contradictory advice. A major purpose of science courses for liberal arts majors is to encourage critical and independent thought in students, to reduce their sensitivity to commercial and political pressures. He was certainly displaying these traits. Not only was his mind racing beyond the confines of the lesson and textbook, but he was challenging the mainstream thought processes and assumptions of astrophysics.

This student did very well in the course. What is more, his objections to conventional thinking regarding dark matter have always stayed with me.

THE PHYSICS OF CONSCIOUSNESS

A second root for the ideas that developed into this book is a chance occurrence much earlier in my professional career. As a graduate student in the early 1970s, I collaborated with a friend, the late Al Fennelly, who was then studying for his Ph.D. from Yeshiva University in New York City.

Al and I were very excited by the concept we had developed for a spacecraft that could reach high velocities and reach the near stars within a human lifetime using cosmic energy fields. We were so happy with the idea that we posted the manuscript to *Science*, one of the most prestigious international peer-reviewed journals.

Following its traditional approach, the journal sent our manuscript to two selected anonymous reviewers. We were saddened to learn from one of these prestigious people that we had neglected some very fundamental physics and merely invented a nonfunctional perpetual motion machine.

The second reviewer was kinder, however, although he also pointed out our error. He identified himself as Evan Harris Walker, who was working at the time at the now-defunct NASA Electronics Center in Boston. Although Harris (as his friends called him) was employed in plasma physics and thus was well suited to review our manuscript, we learned that one of his main interests was quantum physics and its philosophical implications.

Harris instructed us on how we could salvage a portion of our conceptual starship. Even if it could not accelerate to high velocity without fuel, the technique we suggested might have application to deceleration. Following Harris's advice, we revised the manuscript and submitted it successfully to the *Journal of the British Interplanetary Society (JBIS)*, which was beginning a formal study of interstellar concepts. Our 1974 paper eventually evolved into the magsail, a proposed means to decelerate a starship by magnetically reflecting interstellar ions.

Al and I developed a nice professional friendship with Harris, who soon moved from Boston to a position with a U.S. Army research facility in Aberdeen, Maryland. Although the three of us co-authored papers in *JBIS* on in-space propulsion and detecting the planets of other stars, Al and I became fascinated by one of Harris's quantum interests. He had become affiliated with a group of physicists and psychologists who were applying quantum physics to investigate the basis of consciousness and were mostly based at west-coast institutions including Stanford Research Institute (SRI) and Berkeley.

Harris's approach to quantum consciousness was not hard to understand. In his papers and books he discussed the fine structure of neurons—those cells that comprise our brains. Synapses are the structures within neurons that transmit signals from one neuron to another. Synapses are electrically conductive and typically separated by 20 billionths of a meter or 20 nanometers (20 nm). Such a tiny separation between electrical conductors allows for the possibility of various quantum events.

In Harris's theory, wave functions associated with subatomic particles bounce around within the electrical potential wells of the intersynaptic spacing. If you try bouncing a tennis ball against a wall, after a trillion bounces you might conclude that the wall is completely opaque to the ball. But at the quantum level, there is a finite probability that the wave function will instantaneously tunnel though the potential well and materialize somewhere else. This is a very well-established quantum event. Tunneling diodes, in fact, have become standard electronic components.

Initially, Harris and his colleagues were interested in how a thought might move from one part of the brain to another by exploiting quantum tunneling. But some readers realized that a thought could theoretically be transferred not merely from one brain location to another but between brains.

In the depths of the Cold War, this possibility attracted the attention of certain well-funded intelligence-gathering agencies. Imagine that you are a spy. You can certainly lead the glamorous (and likely) short life of James Bond. But how much more comfortable and safer it would be to arrive at your office, perhaps imbibe an appropriate enhancement with your morning coffee, go into a trance state, and efficiently read the minds of your enemies at a safe distance.

Funded by the U.S. Central Intelligence Agency (CIA), the SRI group began to test the alleged abilities of self-described psychics. Al and I learned how one of these, Uri Geller, excelled at the screening tests and used his reputation as a scientific subject to develop a lucrative career on the lecture circuit and as a best-selling author.

When it was demonstrated that Geller's signature cutlery bending could be duplicated by magicians, controversy swirled around his videotaped earlier performance in the double-blind screening tests. Conservatively, Harris presented on one of his visits to New York City a flip chart showing the agreement of the team's second best psychic with the predictions of Harris's theory.

First, he showed a graph representing the theoretical predictions. Second, he flipped a transparency that demonstrated the close approximation of the data points to the theory. Then, with a wry grin, Harris flipped the final transparency showing the error bars. The inherent uncertainty in quantum experiments essentially eliminated the apparent agreement between experiment and theory.

Years later, at a cocktail party I met one of the people claiming to be responsible for uncovering Geller's training as a magician. He demonstrated quite clearly the magic trick of cutlery bending but never explained. As described in subsequent chapters, I have also met some of the physicists who witnessed in person or on videotape Geller's original screening tests. They remain convinced that, no matter how adept he is as a magician, there is no way he could have cheated on these tests. This is the only scientific debate on which I am privileged to be acquainted with honest people on both sides. I suspect that psychic-screening tests could be repeated today with greater accuracy to perhaps resolve this debate.

Harris's enthusiasm and humor were as inspiring as his physics. His equations have been incorporated by C Bangs, who created the chapter frontispiece art in this book, as a form of calligraphy or sacred writing. I would have loved to have discussed the stellar volition evidence considered in this book with Harris, but, sadly, he passed away in 2006. I miss him.

A FAILURE OF NERVE

Another root for the mental activity that evolved into this book was a consulting experience in the early 1990s. I had previously received a bit of attention as co-author with Eugene Mallove of *The Starflight Handbook* (Wiley, NY, 1989), a semi-popular work that began the conversion of interstellar travel from a science fiction subject to a subdivision of applied physics. Since our favored method of exploring nearby interstellar space was the solar photon sail, which is propelled by momentum transferred from reflected sunlight, I was contacted to serve on the science consulting team for a forthcoming science fiction novel.

The novel would be called *Encounter with Tiber*. Since it was to be co-authored by Apollo 11 astronaut Buzz Aldrin, I was overjoyed to join the team. Buzz first requested that I check the calculations for one of the starships in the novel—a sail to be propelled by reflected light from Alpha Centauri A and B, two close stellar neighbors of our Sun.

After confirming to him that the ship and trajectory assumptions made sense, Buzz requested that I take on a more challenging assignment. Since I had received my Ph.D. in the analysis of planetary atmospheres, perhaps I could check one of the plot assumptions. Buzz and his co-author John Barnes assumed the existence of a Jupiter-like giant planet at approximately the Earth's distance from a Sun-like star. I thought that, based upon accepted planetary science, such a planet would evaporate its atmosphere too quickly for life to evolve on any of its satellites.

But one doesn't easily say "no" to Buzz Aldrin, the second person to walk on the Moon. So I promised to check through various source books in my home library to try to locate an equation for the calculation of the lifetime of a planet's atmosphere.

In Chapter 18 of a 1965-vintage graduate level text, *Introduction to Space Science* (ed. W. N. Hess, Gordon & Breach, NY), I located a suitable equation presented by Robert Jastrow and Ichtiaque Rasool, two of my early mentors. I plugged in the numbers for a Jupiter-sized planet close to its star and was amazed to learn that the atmosphere would be stable for billions of years. After checking the results several times, I informed Buzz that his planet's atmosphere would be stable. He checked through my calculations and seemed to be very happy.

But I was not so happy. Could mainstream planetology assumptions be so far off the mark? And, if so, could I successfully challenge them? I mused over the possibility of submitting a short paper to a reputable astronomy journal. But the equation I had used is presented in the sourcebook without derivation. Moreover, I have been unable to locate a similar equation elsewhere in the literature. As an adjunct college professor and consultant, I did not feel strong enough to challenge establishment thought. So I filed my notes and timidly sat on the concept.

A few years later, astronomers in the U.S. and Europe began to discover Jupiter-sized planets circling Sun-like stars. Unexpectedly, many of these are hot Jupiters, giant planets circling well within the Earth's separation from their star.

Author Howard Bloom has written about the responsibilities of a scientist. Even at the risk of his or her life, a scientist must publish his/her conclusions, no matter how heretical they might seem. At least one scientist, Giordano Bruno, was burned at the stake by religious authorities in 1600, for his speculations regarding life on other worlds. I had timidly retreated from one of the cardinal duties of such a scholar. So I promised that if ever I had the opportunity again to challenge an established paradigm that seemed in error, I would advance more boldly. In 2011, I had the chance.

A RETROSPECTIVE IN LONDON

Early in 2011, I received an email from Kelvin Long, who is the current editor of *JBIS*. Kelvin was soliciting participation in a symposium to be held in London in honor of a British science fiction author, Olaf Stapledon. In his 1937 masterwork *Starmaker*, Stapledon had nuclear power, vast astro-engineering projects, and interstellar travel. But the prime focus of this short but influential novel is the conscious evolution of the Universe. Instead of concentrating on one of Stapledon's technological predictions, I elected to investigate in my submission to the symposium one of his philosophical concepts—that stars are volitional and they elect to follow certain paths around the centers of their galaxies.

The first step was to demonstrate that stars, which certainly do not have neurons (the nerve cells in our brains) or microtubules (structures in living cells that some influential microbiologists and physicists believe may contribute to the interaction of consciousness with living matter) might have another avenue to consciousness. But the spectra of cooler red or yellow stars (such as our Sun) have signatures associated with molecules. What is more, it has been known since 1948 that fluctuations in the universal vacuum, from which our Universe arose, affect the bonds that link atoms in molecules. At least one well-known quantum physicist, Bernard Haisch, has speculated that this so-called "Casimir Effect" might be a root of consciousness.

According to this conception, consciousness arises from the interaction of a universal field (which has been dubbed "proto-consciousness") with matter through the agency of fluctuations in the quantum foam that underlies the Universe and is the physical source of creation.

For me to suggest that molecule-rich stars move in part according to a volitional process during a scientific symposium, it would be necessary to find some observational evidence. So, like most researchers in the modern world, I turned on my desktop computer and began a Google search for "stellar kinematics". What I found blew my socks off!

Decades ago, an obscure Soviet era Russian astrophysicist had observed a strange discontinuity in the motions of nearby stars. Dubbed "Parenago's discontinuity" after the discoverer of this anomaly, this affect has since been validated by observations using a European space observatory of about 6000 stellar members of the Milky Way galaxy.

Apparently, cooler, less massive stars move a bit faster in their galactic revolutions that their bluer, hotter colleagues. I realized that the suggested explanation—based upon the rate that infant stars boil off from their birth nebulae—could not be correct because the discontinuity does not occur for very hot, very young stars. Instead, it occurs right at the point in the stellar population distribution where molecules become prevalent in stellar photospheres and upper layers.

Although this discontinuity might be a pointer toward star consciousness, it was still necessary for me to demonstrate a physical method that a conscious star could use to affect its motion. I learned from my research that infant and young stars often eject jets of material. But, in most cases these jets are bipolar, so that jet-caused stellar motion change would be unlikely. Interestingly, I later learned that unidirectional jets emitted from young stars have been observed.

Since I was not aware of these observations at the time of the 2011 Stapledon symposium, I suggested that Evan Harris Walker's original theoretical research on PK should be revisited. A very weak PK (psychokinetic) effect would be ample for a star to alter its galactic velocity by about 20 kilometers per second in the first billion years of its existence—about equivalent to that required by a human to alter his walking speed by less than 1 centimeter per second in a century-long lifespan.

So, I prepared my Stapledon symposium submission to include a discussion of the failure of dark matter searches to find any trace of the stuff, a review of quantum theories of consciousness, and discussions of stellar jets and the PK effect controversy. Because my teaching schedule precluded a trip to London at the time of the Stapledon symposium, Kelvin Long graciously agreed to present my paper.

As expected, it elicited a lot of controversy. To publish it in *JBIS*, I was obliged to submit to the suggestions of four reviewers and shorten it by 50% to appear as an "Invited Commentary." Kelvin suggested that I submit the full version to *Centauri Dreams*, an astronautical/astronomical blog published online by science-journalist Paul Gilster. The blog publication resulted in more than 140 constructive comments and suggestions, some of which are incorporated in this book.

My scientific text was incorporated with C's quantum-referenced art for presentation at a Manhattan artist salon, two science fiction conventions, a joint talk at the London headquarters of the British Interplanetary Society, and an article in the *Baen Press* online science magazine.

Perhaps because of this rapid public attention to the conscious star concept, an alternative explanation for Parenago's discontinuity has been suggested. Called "density waves," this alternative suggests that less massive, cooler stars such as our Sun are sped up in their galactic revolutions as high-density spiral arms of their host galaxy swing by them. Happily (for the conscious star hypothesis), initial observations do not support the density wave concept.

Will the paradigm shift as a result of this concept? Will universal protoconsciousness emerge as a mainstream concept? It is not possible to answer these questions. However, aspects of stellar consciousness are subject to observational validation or falsification. Many of the proposed observational (and experimental) tests of the concept suggested in this book have been suggested by respondents to the *Centauri Dreams* blog entry, for which I am most grateful. As discussed in the following chapters, it may be time for the concept of stellar consciousness to emerge into the mainstream of scientific thought.

FURTHER READING

Dark matter is considered in many college level astronomy textbooks. One popular and successful such text is E. Chaisson and S. McMillan, *Astronomy Today*, 6th edn. (Pearson Addison-Wesley, San Francisco, CA, 2008).

For a popular discussion of Eric Harris Walker's contributions to consciousness physics, see E. H. Walker, *The Physics of Consciousness* (Perseus, Cambridge, MA, 2000). For a more mathematical treatment, see E. H. Walker, "The nature of consciousness," *Mathematical Biosciences*, **7**, 131–178 (1970).

The science fiction book co-authored by Apollo 11 astronaut Buzz Aldrin is B. Aldrin and J. Barnes, *Encounter with Tiber* (Warner, NY, 1996).

An equation that can be used to estimate the lifetime of a planet's atmosphere is Equation (12) on p. 675 of R. Jastrow and I. Rasool, "Planetary atmospheres," Chapter 18 of *Introduction to Space Science* (ed. W. N. Hess, Gordon & Breach, NY, 1965).

The role of the scientist is discussed on pp. 153–154 of H. Bloom, *The Genius of the Beast: A Radical Revision of Capitalism* (Prometheus, Amherst, NY, 2011).

S. Clark, *Extrasolar Planets: The Search for New Worlds* (Wiley, NY, 1998), is one of many sources describing the discovery of hot Jupiters circling other Sun-like stars beginning in the mid-1990s.

My version of Stapledon's 1937 classic is reprinted in O. Stapledon, *Last and First Men* and *Starmaker* (Dover, NY, 1968).

A semi-popular treatment of quantum foam is H. Genz, *Nothingness: The Science of Empty Space* (Perseus, Reading, MA, 1998). A popular treatment of the possible role of the Casimir effect in consciousness authored by a physicist is B. Haisch, *The God Theory: Universes, Zero-Point Fields and What's Behind It All* (Weiser, San Francisco, CA, 2006).

A short version of my contribution to the British Interplanetary Society Olaf Stapledon Retrospective Symposium is published as G. L. Matloff, "Invited commentary—Olaf Stapledon and conscious stars: Philosophy or science?," *JBIS*, **65**, 5–6 (2012). For the longer version, consult Paul Gilster's *Centauri Dreams* blog and search for "conscious stars".

PART I

Ancient nights

Of shamans, myths, megaliths, astrologers, and philosophers

CHAPTER 1

The shaman and the sky

Like to the falling of a star
Or as the flights of eagles are,
Or like the fresh spring's gaudy hue,
Or silver drops of morning dew,
Or like the wind that chafes the flood,
Or bubbles which on water stood:
Even such is man, whose borrowed light
Is straight called in and paid tonight.
The wind blows out, the bubble dies;
The spring entombed in autumn lies;
The dew dries up, the star is shot;
The flight is past—and man forgot.

Henry King, *Of Human Life*

We all belong to a tradition of thought that is, by definition, secular and humanistic. Since at least the Enlightenment, science has tried to steer clear of the ultimate questions: Why are we here? Why was the Universe created? Is there existence after death? And so on.

But this has not always been the case. For most of humanity's 2-million-year sojourn on planet Earth, we did not commute to work, shop at the mall, or entertain ourselves using electronic devices. We did not live in cities or even small villages. Instead, the ancestors of modern humans wandered the planet in small bands surviving initially as scavengers and later as hunter-gatherers. It was an unusual person who traveled more than a few miles from his birthplace, or got to know people outside of his small clan or tribe.

Everything was strange and dangerous. How could we avoid the large predators and hunt the game animals, how sure could we be that the Sun would return in spring and the frozen Earth would thaw? We required a class of people who could intercede with nature and increase our chances of survival. From the dawn of human consciousness until the beginning of settled village existence about 10,000 years ago, the tribal shamans played this role.

Today shamanistic lore survives only in a few isolated Native American and Australian Aboriginal communities and similar settings. Nevertheless, attempts have been made to preserve some of the wisdom that guided the lives of our earliest ancestors.

These people did not write. Record keeping was far in their future. But they did indulge in sympathetic magic to insure that the game beasts would be plentiful and the Sun would rise each morning.

More than 20,000 years ago, brave souls in northern Europe followed their shamans into the deep recesses of caverns. By the flickering light of crude lamps, they created the earliest known art. Mammoth, deer, bison, and other beasts can still be observed cavorting on these cave walls. Some of these early artists dipped their hands in pigment to leave ghostly traces of the creators of these early wonders.

In the same time frame of the High Paleolithic, the greatest cultural flowering of the Old Stone Age, some early humans must have looked at the brightest light in the night sky, our planet's one natural satellite, the Moon. Perhaps they wished to keep track of lunar phases to plan evening hunts, perhaps these were efforts of sympathetic magic to insure the Moon's return, or perhaps these were the work of early women who wished to use lunar phases to keep track of their fertility cycles. We shall never know the motivations of the creators of these artifacts. But what was discovered in the archives of a European museum were bones and antlers of game animals that had been engraved long before the birth of civilization with records of sequential lunar phases.

But what is significant is that the shaman's wisdom offered our ancestors strength and solace in a hostile, unforgiving landscape. We may never know the exact nature of the oral traditions passed along to gathered tribe members by shamans beside the flickering campfires. Unfortunately, writing had not yet been invented and it is almost certain that shamanistic traditions that have survived to modern times have been polluted by the surrounding, more advanced civilizations.

In spite of this unavoidable problem, one cannot help but wonder how these early people viewed the Sun (Figure 1.1) and stars. Were these lights in the sky or were they sentient beings (as hinted at by Gregory Cajete)? As author of this book, I could certainly have reached for a compendium of ancient myths to try to enter the minds of pre-civilized humans to investigate this question. But I reside in Brooklyn, New York, which is evolving to be one of the world's cultural centers. Experts in many diverse fields can be found here. Among them is Donna Henes (also known as Mama Donna), who has adopted the title of Urban Shaman and has published widely in this field. Since she is a fellow artist and among C's professional contacts, I emailed her to arrange an interview to discuss shamanistic traditions.

VISIT TO A SHAMAN

We visited Mama Donna on June 30, 2014 in her spacious loft in the Prospect Heights section of Brooklyn. As a moderate skeptic, I half expected to hear a

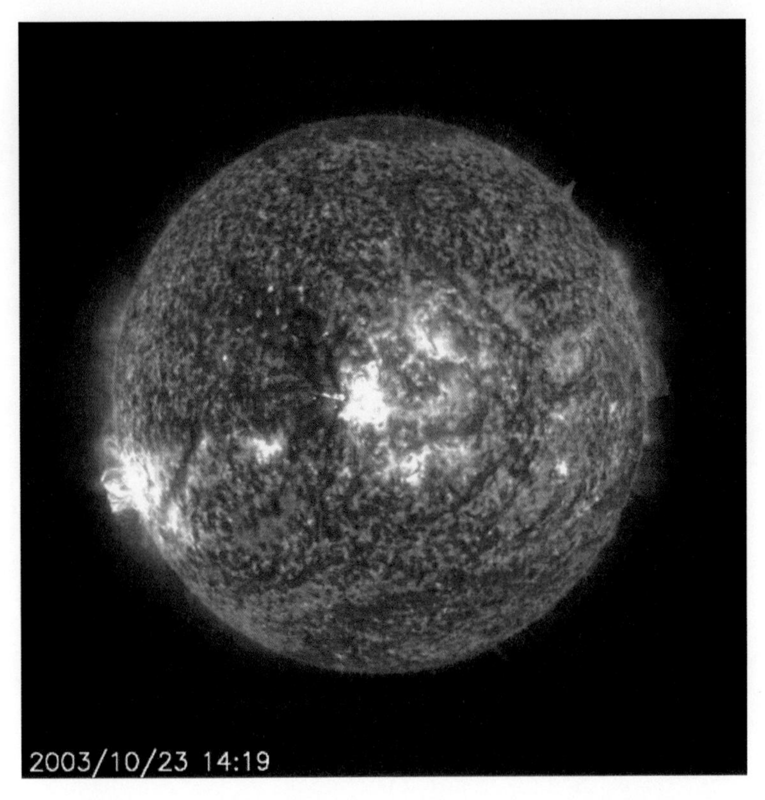

Figure 1.1. Spacecraft image of the Sun (courtesy NASA).

lot of New Age jargon clothed in garments that could never be penetrated by scientific method. I was very pleasantly surprised.

One of the things we discussed was our impression of visits to Neolithic stone circles in the British Isles. Although I had, as an astronomer, been struck by the connection of Stonehenge to the Sun and Moon (discussed in Chapter 2), Donna had been affected by the apparent connection of earlier megalithic monuments in the U.K. (especially Silbury Hill and Callanish) to the Earth. When she described her reaction to the perceived energy of the standing stones, she also discussed how those who accompanied her corroborated her perceptions.

Shamanism sprouts from a time when human social structures were quite egalitarian. Organized warfare had not been invented. The concept of class and gender differentiation was still in the future. Many people today scoff at shamanistic rites. But in the days before organized medicine, lore regarding beneficial and harmful herbs was passed from one generation to the next by the Wise Women of the tribe. More than a few of these matriarchs were burned as witches by the authorities of organized Christian religion during the Middle Ages.

The shaman has a different concept of time than is subscribed to by modern Westerners. Most people today accept the concept of linear time and progress. Our lives are better than those of our parents. Furthermore, our parents were happier (and had more possessions) than our grandparents. To a shaman, time is cyclic. Things repeat, probably in accordance with the dance of the seasons. Probably, shamans first observed celestial objects as a means of keeping track of terrestrial cycles.

We also have a possibly skewed vision of our place in the cosmic scheme of things. Humans are at the crown of creation, followed in turn by higher animals, insects, plants, and things of an inanimate nature. To a shaman, everything is alive and conscious including trees and even rocks.

When I asked Donna about a shaman's approach to the concept of stellar consciousness, she responded by reminding me of Carl Sagan's famous quote that we are all "star stuff." Many of the atoms that comprise our bodies have been recycled many times through the biosphere, but they originated in the interior of a dying star about 5 billion years ago. Perhaps the consciousness of the birth star has been passed along to the atoms expelled from its ashes. Another interesting point is that, in our scientific age, we have no real definition of life and death. Stars die. And to quote Donna, "how can a star die if it had not lived?"

The modern practice of shamanism includes rituals, chants, channeling, goddess worship, and interpretation of past life experiences. Much of this material is beyond the bounds of scientific investigation, at least at the present time. But some shamanistic practices—including recognition of the wisdom of children and direct, mindful connection with ancestors—might be of great value to our present civilization.

It occurred to me after the interview that shamanism in its modern form might have a lot to offer in consideration of the ultimate questions. Perhaps life and death are complementary aspects of the same thing, in a not dissimilar way to the complementary aspects of waves and particles in the physicists' description of photons and electrons.

Before we concluded the interview, we exchanged books we had authored. One of the books authored by Donna is *Celestially Auspicious Occasions: Seasons, Cycles, and Celebrations*. Two of the people who assisted Donna with this book are astronomers: Susie Chippendale, director of the Santa Fe Community College Planetarium and Neil de Grasse Tyson, director of the Hayden Planetarium in New York City. The next section "Solar tales" discusses shamanistic lore from this source and the following section "Sky lights and a river in Heaven" deals directly with stellar and solar consciousness.

SOLAR TALES

Pre-literate people recognized their dependence on the Sun and closely identified with it. Though most recognized the Sun as a god and others as a goddess, organized solar worship and ritual was likely far in the future of early shamanistic people. But because some nomadic Paleolithic societies have survived until modern times, we have some knowledge of solar myth.

One such tribe is the Havasupai who once populated a portion of Arizona near the Grand Canyon. Donna Henes reproduces the Havasupai *Prayer to the Sun*:

> Sun, my relative, be good coming out.
> Do something good for us.
>
> Make me work,
> So I can do anything I wish in the garden.
> I hoe, I plant corn, I irrigate.
>
> You, sun, be good going down at sunset.
> We lie down to sleep. I want to feel good
> While I sleep, you come up.
>
> Go on your course many times.
> Make good things for us men.
> Make me always the same as I am now.

To other early people, the Sun represented the Eye of God. This is evidenced in the traditional Gaelic *Song to the Sun, Ortha nan Gaidheal*, which also is transcribed by Donna Henes:

> Thou eye of the Great God
> Thou eye of the God of Glory
> Thou eye of the King of creation
> Thou eye of the Light of the living
> Pouring on us at each time
> Pouring on us gently, generously
> Glory to thee thou glorious sun
> Glory to thee thou Face of the God of life.

Equally powerful are the myths presenting the Sun as a goddess or the garment of a goddess. In Vedic tradition, the Sun is a golden garment adorning the Great Goddess. The Buddhist sun goddess Mari may be a precursor to the Christian Mother of God, Mary.

To these ancients, the Sun was many things. It was a protector and a bringer of light. But it was also a benevolent friend who could be depended upon. It was a good deal more than a ball of hot gas hanging in the firmament at a distance of 93 million miles or 150 million kilometers.

SKY LIGHTS AND A RIVER IN HEAVEN

Today, we know that those steadfast lights in the night sky called stars are actually very distant suns, and that curious cloud-like band crossing the sky called the Milky Way (Figure 1.2) is a spiral arm of our home galaxy, which is made up of billions of stars. But before the invention of the telescope a few centuries ago, myth took the place of science in describing these celestial features.

In all likelihood, the most ancient human culture on Earth is that of the Australian Aborigines. According to E. C. Krupp, the Euahlayi people of Australia considered the Milky Way to be a path leading to the dwelling of the sky gods. The Yakuts of Siberia echo this approach in describing the Milky Way as "God's footprints." In another Yakut interpretation, the Milky Way is actually an elixir dispensed to the dying by the Mother Goddess.

New World indigenous people had their own take on the Milky Way. The Osage people of Kansas and Missouri believed that the souls of the dead traverse the Milky Way until they reach a convenient star to reside in. To the Skidi Pawnee, the Milky Way was traversed by departed souls on their way toward their ultimate celestial home, the Southern Star.

The Passamaquoddy people of northeastern North America composed Song of the Stars, which has been transcribed by Donna Henes:

> We are the stars which sing
> we sing with our light
> we are the birds of fire
> we fly over the sky
> our light is a voice
> we make a road for spirits,
> a road for the Great Spirit,
> amongst us are three hunters
> who chase a bear
> there never was a time when they were not hunting,
> we look down on the mountains
> this is the song of the stars.

Figure 1.2. COBE spacecraft image of the Milky Way (courtesy NASA).

In these poignant verses are captured the beliefs of these people that the stars are conscious and immortal and connected in some way with the fate of humans.

Modern humans are further along the spiral of development and intellectual sophistication than the ancient Passamaquoddy. We certainly have a better knowledge of the dimensions of the Universe and the duration of deep time. But wouldn't it be nice if we could somehow recapture some of their wonder in confronting this grand Universe that we inhabit?

FURTHER READING

An excellent source describing the Paleolithic lunar engravings on animal bones and antlers is A. Marshak, *The Roots of Civilization* (McGraw-Hill, NY, 1972). Some of the solar and stellar shamanistic myths discussed above are from D. Henes, *Celestially Auspicious Occasions: Seasons, Cycles, and Celebration* (Perigee, NY, 1996). Another source of sky myths is E. C. Krupp, *Beyond the Blue Horizon: Myths and Legends of the Sun, Moon, Stars, and Planets* (Harper-Collins, NY, 1991).

Consult Donna Henes's website *http://www.donnahenes.com* for additional information on the practice of modern shamanism and a list of her many publications.

An interesting discussion of tribal science is G. Cajete, *Native Science: Natural Laws of Interdependence* (Clear Light Publishers, Santa Fe, NM, 2000). In this, he mentions that among the Skidi Pawnee, stars were considered to be sacred beings who desired relationships with their human relatives.

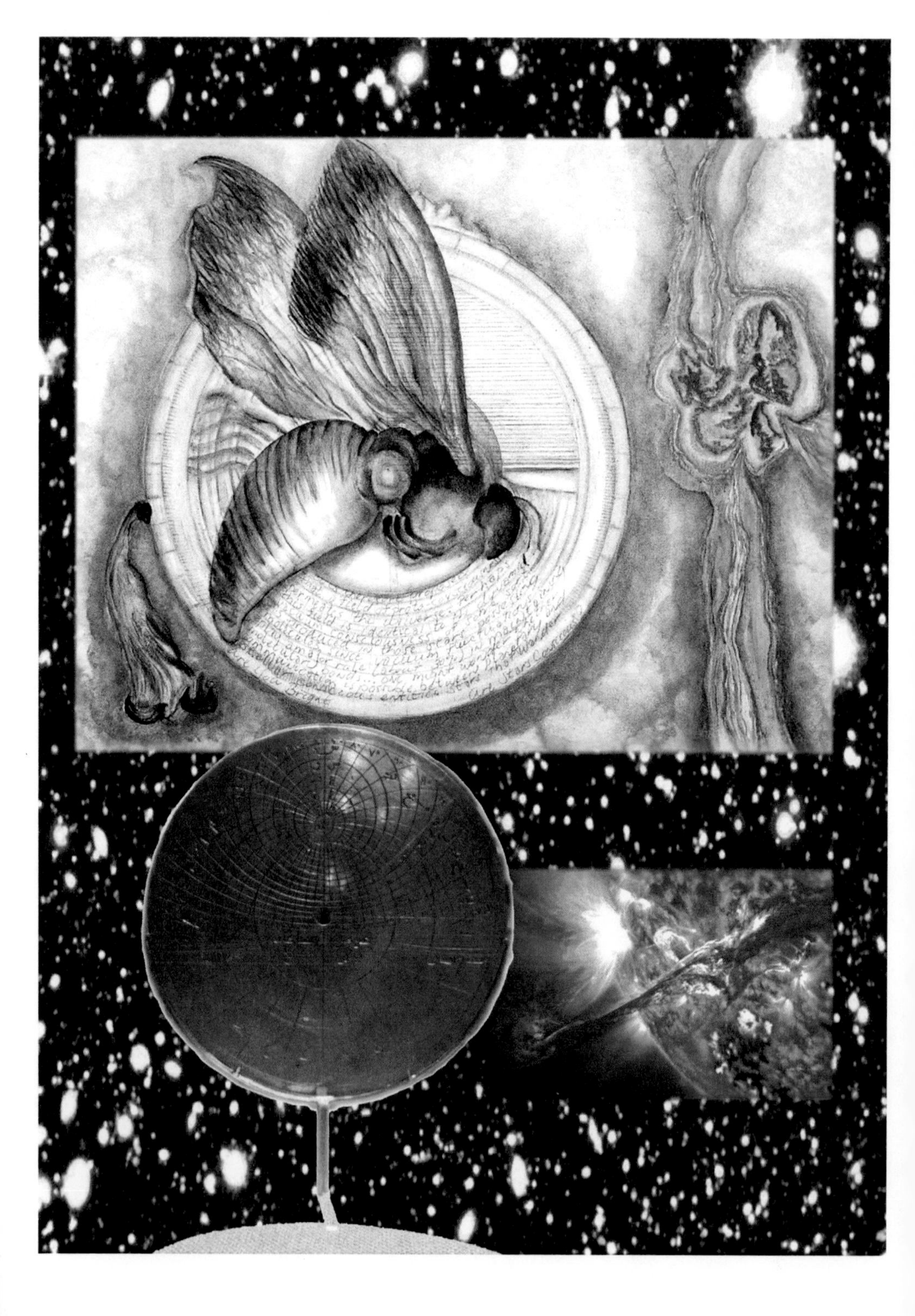

CHAPTER 2

Earth Mother, Sky Father

I saw Eternity the other night,
Like a great Ring of pure and endless light,
  All calm, as it was bright:
And round beneath it, Time in hours, days, years,
  Driven by the spheres
Like a vast shadow moved; in which the world
And all her train were hurled.

Henry Vaughan, *The World*

After perhaps existing as nomadic hunter-gatherers for 2 million year, our ancestors began to experience a vast change in lifestyle about 10,000 years ago. For the first time, humans constructed permanent settlements—towns such as Catal Huyuk in what is now Turkey and Jericho in the Middle East. No longer would we nomadically follow the herds and depend upon our hunting prowess for survival. Perhaps triggered by climate change, this transition from the Paleolithic (Old Stone Age) to the Neolithic (New Stone Age) may have been the greatest lifestyle modification thus far in human history or prehistory.

GOD AND GODDESS

Agriculture and animal husbandry were developed, followed by trade between communities and the beginnings of record keeping. Some of the surviving art from this period seems to present the beginnings of organized warfare. Perhaps it was necessary for comparatively comfortable village and agricultural communities to defend themselves from predation by nomadic neighbors.

Certainly, human psychology changed with the cultural environment. The earliest religions might date from these times and there were in all likelihood two deities. Sky Father was on top—it was his job to produce solar illumination and precipitation in the heavens. The function of this action was to fertilize Mother Earth.

Both Sky Father and Mother Earth were powerful and inhuman—but they could be appeased by appropriate rites and sacrifice. It was not unknown for humans to be sacrificed in this period. Perhaps the story of Abraham and Isaac in the Bible represents a rejection of the concept that the Earth and Sky deities required human sacrifice. This tale may have been orally passed along for millennia before it was finally transcribed.

As the Neolithic progressed, architects honed their skill with stone and brick. In parts of the civilized world, huge stone monuments or monoliths were erected. Although, there are no written records from this time, it is possible that Sky Father and Earth Mother were diversifying into additional entities. One celestial male deity may have been identified with the Sun. A celestial female may have become the Moon because of similarity in the duration of women's menstrual cycles and the lunar phase cycle. While Earth Mother still ruled the solid surface of our planet and insured the seasonal continuity necessary for agriculture, a subsurface Earth God may have been identified with earthquakes, volcanoes, and tsunamis.

CIRCLES OF STONE

Before the invention of the first scripts, the construction of the Egyptian pyramids, and the start of the Bronze Age, residents of the Mediterranean island of Malta raised megalithic temples to the Earth Mother which pioneered the use of the trilithon. The trilithon is a free-standing structure composed of two large, vertical stones capped by a horizontal stone serving as a lintel.

During the late Neolithic, this culture may have radiated into the British Isles and merged with the local inhabitants. The construction of the most famous megalithic monument in the U.K., Stonehenge, began about the same time as the pyramids, and made extensive use of the trilithon. Figure 2.1 presents a photograph, from a NASA website, that shows the current state of the monument. Some of the stones in this structure have an estimated weight in excess of 30 tons (30,000 kg). Quarrying, transporting, dressing, and raising them must have been quite a challenge for Neolithic people. In its original state, Stonehenge must have been most impressive. Its now nearly obliterated outer bank had a diameter in excess of 100 meters (300 feet).

Writing was perhaps a millennium in the future at the start of the megalithic era; the discipline of historical record keeping was even more distant in time. So trying to determine the mindset of the builders of Stonehenge and other megalithic monuments is a challenge.

However, some astronomers and antiquarians have risen to this challenge. Working under the auspices of the Glasgow Parks Department, a team including Duncan Lunan and Archie Thom erected Sighthill in the late 1970s, the first Scottish stone monument in more than 3000 years designed to view celestial alignments.

Although traditions told of the celestial alignments at Stonehenge, it was not until the dawn of the computer age that scientists could rigorously study these alignments. Two researchers who have investigated the astronomical

Figure 2.1. Stonehenge today (courtesy NASA).

significance of this site are the British astrophysicist Fred Hoyle and the British American astronomer Gerald S. Hawkins.

Rather than using its current somewhat decayed configuration, Hawkins' strategy was to input the plan of Stonehenge in its un-ruined original state into the Boston University computer and then input the positions of various celestial objects as they would appear about 4000 years ago. Imagining that he was standing at the so-called "altar stone," at the monument's center, he searched for the possibility of sighting certain celestial objects through the gaps between the two vertical components of various trilithons.

As reported by Hawkins and other researchers, most alignments were for the Sun and Moon. Figure 2.2 is a NASA reconstruction of a summer solstice alignment at Stonehenge in about 2400 BC.

Whatever the motivations of Stonehenge's architects and designers, it is unlikely that the purpose of the monument was scientific. Most archeologists would probably agree that the functions of Stonehenge and similar stone circles were primarily religious and ceremonial. The principal deities involved with the site must have been early versions of the Sun god—who would later evolve into Apollo or Hermes—and a Moon goddess—who would later be called Diana or Artemis.

Hawkins speculates that one function of Stonehenge was to allow the priest-astronomer class to maintain power in the community by predicting lunar eclipses. There are 56 chalk-filled so-called Aubrey holes (named for John Aubrey, the discoverer) arranged around the circumference of the earliest version of the monument. It is not impossible (although it is certainly not

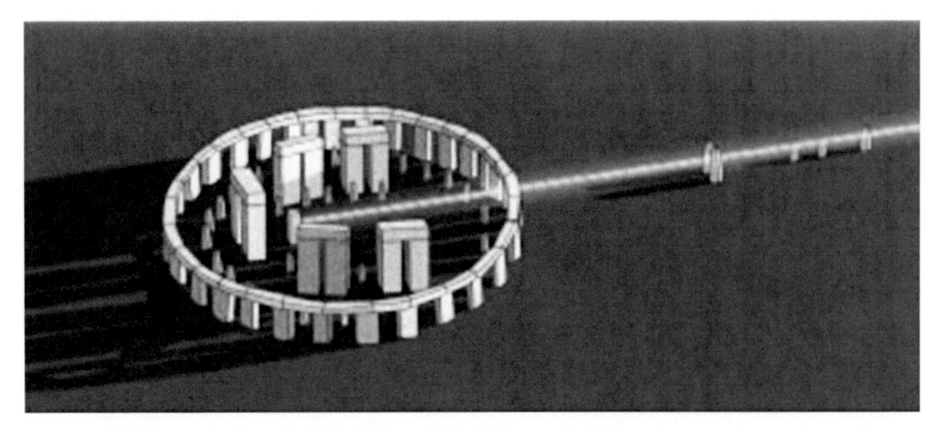

Figure 2.2. Summer solstice at Stonehenge, 2400 BC (courtesy NASA).

universally accepted) that priest-astronomers could keep track of the lunar eclipse cycle by moving a stone once a year between adjacent Aubrey holes.

In this manner, priest-astronomers using Stonehenge could alert the local populace of a forthcoming lunar eclipse. When viewed through unpolluted, dust-free skies, the eclipsed Moon often turns blood red as a consequence of reflected sunlight that has been refracted through Earth's atmosphere. The result, a bleeding Moon goddess (Figure 2.3), would have been a fearsome sight to Neolithic people. Various rites and sacrifices might have been required to save the Moon from the attack of a celestial monster or dragon.

Stonehenge, because of its location and state of preservation, is the most famous and best studied stone circle. But it is certainly not unique.

George Dimitriades of the University of Genoa has examined the distribution of stone circles and other megalithic monuments in Europe. There may be a correlation between the locations of these Neolithic and Bronze Age structures and deposits of mineral resources, including silver, gold, and copper. There may be a mythological connection between certain standing stones and the shape of the human body. It is not impossible that some of these stones represent ancestors of the local people. Some very ancient myths such as the labors of Hercules may have originated from the construction of megaliths.

The earliest known stone circle predates Stonehenge by at least a millennium. Its ruins are located at Nabta Playa in Nubia (modern day northern Sudan and southern Egypt) and may be about 7000 years old. With the application of satellite imagery, it is argued by Thomas G. Brophy that ancient pre-Egyptian people may have used this monument to observe stellar alignments rather than lunar and solar positions. The stars in question may be the three belt stars in the constellation Orion and the alignment in question may involve the appearance of these stars on the local meridian (the imaginary line in the sky through the north and south directions and the overhead point or zenith) before sunrise on the midsummer solstice. The Egyptian harvest god Osiris became identified with the constellation Orion.

Figure 2.3. The Moon seems to bleed when eclipsed (courtesy NASA).

Were this stone circle 1000 years older than its estimated age, it could have been used as a calendar circle to keep track of Sirius, the brightest star in the sky. When considering ancient people's thoughts about stellar consciousness, it is interesting that the Egyptian goddess Isis later became identified with Sirius.

More recent megalithic monuments and similar wooden structures have been found in the New World. But, as noted by Gerald S. Hawkins, lack of written records, associated surviving myths, and poor states of preservation make it difficult to ascertain the function of these structures.

MOON MAGIC

Not many sky traditions from this period have survived to our era. But there are a few. According to David and Carol Allen, the first astronomers we know by name were two Bronze Age Chinese brothers named Ho and Hsi. Using a stone circle (or some other applicable device), they informed the local ruler of a forthcoming lunar eclipse. The ruler prepared for the required ceremony. There must have been priests, soothsayers, acrobats, dancers, and magicians.

But, at least early on, eclipse prediction must have been an inexact art. When the eclipse failed to appear as scheduled and the local ruler suffered severe loss of face, Ho and Hsi were executed. In that era and location at least, certain calculation eras were a capital offense!

FURTHER READING

A good source for the history of the ancient world is the classic by J. Hawkes, *The Atlas of Early Man* (St. Martin's Press, NY, 1976). The saga of Sighthill is discussed by D. Lunan, *The Stones and the Stars* (Springer, NY, 2013).

There are many sources describing the astronomical alignments of megalithic monuments. My favorites are G. S. Hawkins, *Stonehenge Decoded* (Doubleday, Garden City, NY, 1965) and F. Hoyle, *On Stonehenge* (W. H. Freeman, San Francisco, CA, 1977).

Many British antiquarians active in the 19th century and earlier were fascinated by Stonehenge and other U.K. megalithic monuments. An updated version of one originally written in 1880 by W. M. Flinders Petrie is G. S. Hawkins, *Stonehenge: Plans, Description, and Theories* (Histories & Mysteries of Man Ltd, London, UK, 1989).

If you are fascinated by lunar eclipse lore, there are many sources available. One is D. Allen and C. Allen, *Eclipse* (Allen & Unwin, Boston, MA, 1987).

A comprehensive study of the distribution of European megaliths is G. Dimitriades, "Looking for metals: Megalithic monuments between reality and mythology," *Proceedings of the October 29–30, 2008 Sofia International Conference Geoarchaeology and Archaeomineralogy* (ed. R. I. Kostov, B. Gaydarska, and M. Gurova, St. Ivan Rilski Publishing House, Sofia, Bulgaria, 2008, pp. 205–210).

Many scholars have investigated celestial alignments at the Nabta Playa stone circle. See T. G. Brophy and P. A. Rosen, "Satellite imagery measures of the astronomically aligned megaliths at Nabta Playa," *Mediterranean Archaeology and Archaeometry*, **5**, 15–24 (2005).

Egyptian theology and mythology are connected with various celestial objects, as described in E. C. Krupp, *Beyond the Blue Horizon* (HarperCollins, NY, 1991)

New World megaliths are described in G. S. Hawkins, *Beyond Stonehenge* (Harper & Row, NY, 1973).

CHAPTER 3

Wandering sky gods

Each night to the towering summit of the sky—
The one who did all that came journeying by;
And strange though it may sound, it was easy to see
That it could have been none other, that it was he.

Merrill Moore, *Constellation*

It's 1000 BC, roughly 3000 years ago. Things have changed in the civilized world. The Bronze Age is fading, to be replaced by the time of Iron. Armies strive across portions of the Near East and the Aegean. Trojans and Greeks have indulged in their civil war. Hebrews and Philistines strive for control of a small section of land bordering the Mediterranean Sea. Aryans now control the Indus Valley and the caste system is emerging in India.

The alphabet has been invented and is beginning to replace earlier scripts. The scribe class will become obsolete and literacy will slowly spread. Restless human groups—Celts, Etruscans, and others—migrate across great distances, sometimes threatening the established order in the lands they enter.

In what will someday be called the United Kingdom, Stonehenge is a ruin. Its ancient purpose, lost with its builders in the mists of prehistory, has been replaced by the rites of Druids, the priestly caste of the local Celts.

But civilization continues through this time of turmoil and change. In what is now Iraq and Iran, the people of Babylon have erected multi-story stone towers called ziggurats. From the tops of these structures, a new class of priest-astronomer continues the exploration of the heavens. Unlike their Neolithic and Bronze Age forebears, they do not concentrate their efforts on the primary lights in the day and night sky—the Sun and Moon. Instead, they carefully observe the shifting positions of those wandering star-like objects that they call "planets."

To these sky observers, Earth of course was not a planet, it was humanity's home base and distinct from the celestial realm. However, the naked-eye planets Mercury, Venus, Mars, Jupiter, and Saturn were the targets of these early astronomers. Their observations of planetary positions would lead to the later Greek models of the Solar System, or cosmogonies.

Table 3.1. Naked-eye planet names according to three ancient traditions.

Roman	*Greek*	*Babylonian*
Mercury	Hermes	Nabu
Venus	Aphrodite	Ishtar
Mars	Ares	Nergal
Jupiter	Zeus	Marduk
Saturn	Kronos	Ninib

Table 3.1 gives the names of the naked-eye planets according to the Romans, Greeks, and Babylonians. Today, we associate Roman names with the naked-eye planets. It is significant that, in all three traditions, the naked-eye planets or wandering stars are named for major gods and goddesses in the ancient pantheons. Nabu was the god of wisdom and a patron of scribes, Ishtar was a love and fertility goddess, Nergal was a war god (probably because of the planet's red color), Marduk was the king of the gods, and Ninib was a former Sun god now deposed to represent a modest planet.

It is likely no coincidence that the naked-eye planets are named for various classical and pre-classical gods and goddesses. In a victory for the art of marketing, the priest-astronomers convinced many influential residents of ancient city states and empires that these moving lights in the sky actually were divine.

This identification of planets with divine entities was not limited to ancient Babylon, Greece, and Rome. At about the same time in human history, similar approaches to viewing the motions of Solar System bodies were being adopted in Egypt, India, and China. The brightest of the naked-eye planets, Venus, would also become important in the religious thought of New World people including the Aztecs and Mayas of Central America.

Even though we live in a secular age, the days of the week are named for moving lights in the sky. Sunday, of course, belongs to the Sun and Monday is the Moon's day. Tuesday, Wednesday, Thursday, Friday, and Saturday are respectively identified with Mars, Mercury, Jupiter, Venus, and Saturn.

Perhaps because of the ancient popularity of this celestial identity system, Babylonian priest-astronomers morphed into the first astrologers. If you had the resources, you could commission one of these authorities to cast your horoscope. Your personal characteristics might have a great deal to do with planet positions in the sky on your day of birth. You could plan your daily activities using this sky guide.

In the classical world, many rulers would not sign a treaty (or break one) without first consulting the opinion of Jupiter. A general preparing for a significant battle would be very interested in the celestial location of Mars. If you were planning a business deal or a trade arrangement, Mercury and Saturn could be consulted for advice. And if you were a Lady of the Court interested in conducting a love affair, it certainly might help to follow the advice offered by Venus.

Today astrology is, at least to most educated people, a parlor game. Perhaps the most famous pick-up line at a cocktail party is "what's your sign?" But in the ancient world, the position of the Sun among the 12 constellations of the zodiac on your date of birth was of great significance. This cannot be said of all ancient people, since astrology was satirized by ancient Hebrew authors in the Biblical tale of the "Tower of Babel."

SKY MYTHS

Long before the dawn of electricity, these civilized and literate people tried to make sense of the night sky literally filled with stars. They were a long way from understanding these distant lights that appeared and vanished in accord with diurnal and seasonal cycles. Perhaps to become more comfortable with this celestial immensity, they filled the sky with heroes, gods, and demons. In some cases, divine and mortal consciousness was identified with stellar hosts.

Sky twins

Consider the constellation Gemini, for example. The two brightest stars in this zodiacal constellation are Castor and Pollux. Perhaps these stars are of similar visual brightness and are near neighbors (only 4.5 degrees apart) on the celestial sphere. They were identified as the Great Twins by the Sumerians.

To the Classical Greeks and Romans, these stars were the twin children of Zeus (the ruler of the Olympian gods) and Leda, a married mortal woman seduced by Zeus while disguised as a swan. Both boys hatched from the same egg and became heroes, joining with Jason on the quest for the Golden Fleece. Unlike his immortal twin, Castor was mortal. After Castor was killed in a fight, his brother beseeched Zeus to bring him back to life. Ultimately, these two now-immortal twins were transferred to the sky as Alpha and Beta Gemini, the two brightest stars in that constellation. As well as eternally sharing each others company in the celestial realm, the rising of these stars before sunrise in late spring heralded calm summer seas and was welcomed by sailors.

Other ancient people also viewed these stars as conscious entities. Although Arabian astronomers also called them The Twins, some medieval Arabian sky maps picture them as a pair of peacocks. In ancient India they were identified as The Horsemen.

Although no longer considered to be inhabited by divine mortals, Castor and Pollux are fascinating astronomical objects. Castor is the 23rd brightest star in the sky, with a visual magnitude of 1.59. A blue-white hydrogen-burning main sequence star, it is at a distance of about 45 light years and actually consists of three components, each of which is a binary star. Pollux is an orange-red giant at a distance of about 35 light years and has a visual brightness slightly greater than its twin.

Star-crossed lovers

Along with Deneb, the bright stars Altair and Vega form the Summer Triangle, a prominent star grouping in the northern hemisphere summer sky. Between these two stars is the Milky Way.

Ancient Chinese sky watchers had no idea that the faint, nebulous substance of the Milky Way is actually the combined radiance of billions of faint stars. Instead, they believed it to be a celestial river. Around 700 BC, a myth developed around these stars and the "river" that separates them.

Altair is the Cowherd and Vega is the Weaving Lady. Although these mortal lovers were originally reunited in heaven, an early Chinese deity, the Queen Mother of the West, became annoyed by the lovers' celestial activities and drew a line between them. This line became the Milky Way, a celestial river that always separates Altair and Vega.

Seven Sisters—or maybe only six

The most famous mortals to be identified with stars are the ladies of the Pleiades (Figure 3.1). Viewed under low magnification through binoculars or a small telescope, a few hundred young stars might be counted in this open cluster, which is classified as M45 in the catalog of the 18th century French

Figure 3.1. The Pleiades as seen by the Hubble Space Telescope (courtesy NASA).

astronomer Charles Messier. On a clear night far from city lights, an observer with excellent eyesight is lucky to count six visible Pleiades. So it is somewhat of a mystery that so many ancient traditions refer to this grouping as the Seven Sisters. I don't think that people on average had better eyesight in antiquity than today. So we are left with two alternatives: either one of the brightest of the young, evolving stars in this grouping has decreased in brightness a bit over the last few millennia or Earth's skies are a little less transparent than they used to be.

But not all ancient people saw the Pleiades as seven stars. Some Chinese, Japanese, Aboriginal, and Native American people recognized six women associated with the bright, young, massive stars in this grouping rather than seven.

The most famous myth regarding the Pleiades is the Greek story of a pursuit of seven maidens by the sexually aroused hunter Orion (who is also a sky resident). In answer to their pleas, Zeus placed both pursuer and pursued in the sky. In the celestial realm, the pursuit continues. In some traditions, the seven maidens were companions of the goddess Artemis.

In a Hindu myth of the 5th century BC, there is a connection between the Pleiades and the stars of the Big Dipper. In this myth the seven visible stars in the Big Dipper are sages and the stars of the Pleiades are their wives.

To ancient Incan astronomers, the Pleiades were not individual entities. Instead, they were the eyes of Viracoha, the god of thunder.

In the vicinity of the zodiacal constellation Taurus the Bull there is another open cluster, the Hyades (Figure 3.2). This is a more mature open cluster than the Pleiades and the nebulosity of the stellar birth nebula has dispersed. According to Greek myth, the visible members of this cluster were daughters of Atlas, the titan responsible for holding up the sky.

How the Sun God gets around

Some sky myths attempt to explain how the Sun God moves through the sky. Ancient Egyptians believed that the Sun God Ra rose in the east and sailed in a solar boat across his celestial realm before setting in the west. According to Greek sources, Helios (also referred to as Apollo) was transported instead in a chariot drawn through the heavens by four celestial steeds. As an example of how powerful such imagery is, even to scientifically sophisticated moderns, Figure 3.3 presents the mission patch of the aborted Apollo 13 lunar expedition. Note there are only three celestial steeds pulling the Sun in this image. Where is the fourth?

Figure 3.2. The Hyades as seen by the Hubble Space Telescope (courtesy NASA).

Figure 3.3. Apollo 13 mission patch (courtesy NASA).

FURTHER READING

An excellent source for ancient sky legends, history, and myth, is E. C. Krupp, *Beyond the Blue Horizon* (HarperCollins, NY, 1991). Krupp directed at that time the Griffith Observatory in Los Angeles. Another source of sky myths and legends is T. Condo, *Star Myths of the Greeks and Romans: A Sourcebook* (Phanes Press, Grand Rapids, MI, 1997). A third source is A. Aveni, *Sky Watching* (University of Texas Press, Austin, TX, 2001).

Bright star data and some related myths can be found in the three-volume compendium by R. Burnham, Jr., *Burnham's Celestial Handbook:An Observer's Guide to the Universe beyond the Solar System* (Dover, NY, 1978).

Information on the Greek solar deities Apollo and Helios is available online (*http://wikipedia.org/wiki/Helios*).

CHAPTER 4

Medieval mysteries

> Autumn hath all the summer's fruitful treasure;
> Gone is our sport, fled is our Croydon's pleasure.
> Short days, long nights come on apace;
> Ah, who shall hide us from winter's face?
> And here we lie, God knows with little ease.
> From winter, plague and pestilence, good Lord, Deliver us!
>
> Thomas Nashe, *Autumn*

It's the year 415 AD. To many observers, the events that transpired in sunny Alexandria must have seemed to foretell the most severe winter that one could imagine. This was indeed the autumn of the classical world. Bands of savage barbarians wandered the formerly safe streets of ancient cities; the new belief of Christianity warred with pagan humanism. All was in flux through the great Roman Empire.

In the beautiful city of Alexandria, which was named in honor of Alexander the Great, the Greek conqueror who had unified much of north Africa and west Asia during the 4th century BC, stood a magnificent library. The Royal Library of Alexandria had, for most of a millennium, served as the foremost research center and knowledge depository of the ancient world.

In the doorway of the Library stood an accomplished and beautiful woman. Hypatia, herself the daughter of a noted philosopher, was the last director of the Royal Library and the earliest female astronomer whose name has been preserved. She was chaste, a member of an order similar to the Vestal Virgins, an organization that later was folded into Christianity as an order of nuns.

However, this day there was to be no peaceful incorporation into the new religion. A torch-bearing mob of enraged Christian fundamentalists approached the Library. This sacred procession was led by a future saint, Cyril, the Archbishop of the city.

Hypatia must have realized the futility of continuing her work and the impossibility of stopping these fanatical citizens. Before they torched the Library, they dragged her into the street. They stripped her of her clothes. Anyone familiar with the way religious mobs treat women must wonder how long her chastity was intact.

After having their way with her, these upright citizens flayed the flesh from her body. Hypatia's death must have been long and painful. Her earthly remains were consumed by fire and her works were lost.

Not much of the vast treasure of classical literature and knowledge survived the conflagration. Only 7 of the 123 plays of Sophocles, for example, have come down the ages intact.

The Empire would survive for a few more decades, but this murder of an early scientist served as the death knell of free classical thought, at least in the West for about a millennium. If a Pope claimed that the Moon was made of blue cheese, the only safe recourse for a prudent student of the skies was to investigate what variety of blue cheese would fit the bill. It is interesting to note that the religiously orchestrated murder of another scientist—Giordano Bruno—who was burned at the stake by Roman Catholic authorities in 1600, denotes the re-emergence of critical thought in the Western world.

Were it not for the illuminations and copies of a few classical manuscripts produced in Islamic mosques and Irish monasteries, the traditions of civilization might have perished with Hypatia. But some ancient knowledge was preserved in the strange writings of medieval mystics.

MYSTERY CULTS

There have always been mystery cults with select memberships which preserve and enlarge upon occult knowledge. Such cults must have proliferated in the waning days of the Western Roman Empire, when Paganism, Christianity, and many other doctrines vied for adherents and recognition. One such cult was Hermeticism, which is traditionally considered to be based upon the teachings of an ancient philosopher, Hermes Trismegistus. Hermetic concepts apparently influenced early Christian figures including Thomas of Aquinas and Augustine, Renaissance scientists including Giordano Bruno and Isaac Newton, and later transcendentalists such as Emerson. Poets including Yeats have been influenced by such writings.

Hermeticism

At least some of the wisdom of antiquity—gathered from the discourse of deductive philosophers—seems to have been preserved in altered form in Hermetic doctrine. This doctrine seems to be a reaction both to pure rationalism and doctrinal faith.

One of the theological precepts of Hermeticism is that there is one god and that this transcendent, unitary entity stands apart from but participates in the material universe. The ancient mysteries of birth and death are contained in Hermetic thought as is the Christian concept of resurrection. There are three basic and interconnected levels to perceived reality: physical, emotional, and mental.

Although the ancient art of alchemy is included in Hermeticism, it is other than the familiar act of transmuting lead to gold. Alchemy is identified with the Sun and is concerned with the basic philosophical mysteries as well as such chemical processes as distillation and fermentation.

Astrology is not only the relation of divine planets to mortal life but also a representation of the operation of the stars. Humans still have free will in Hermetic thought. Although celestial events influence our lives, our actions at least partially dictate events.

The Hermetic view of the Universe is based the Ptolemaic or geocentric Solar System model. The Earth is at the center, circled by seven heavens which are the orbits of the naked-eye planets, the Moon, and the Sun. This was the accepted Solar System model until the Renaissance, when it began to be replaced by the modern heliocentric or Sun-centered model.

Many scholars view Hermeticism as a tolerant philosophical religion that is flexible and moderate and considers mind to be omnipresent in the Cosmos. In that respect at least, it certainly is in line with the concepts developed in this book. Although Hermeticism is significant, it is not unique.

Neoplatonism

A related ancient cult that survived the fall of Rome is Neoplatonism, which is derived from the precepts of the great Athenian philosopher of the 5th century BC, Plato. Some classical scholars including Plutarch are considered to be Neoplatonists. Some doctrines of this school of philosophy were adopted by early Church fathers.

Neoplatonism divided the world into visible and invisible parts. Falls (into sensual and depraved habits) were introduced as the cause of the soul's separation from the eternal divine.

But this doctrine is essentially pantheistic. Divinity is not considered to be entirely separate from the world. In a sense, everything is divine.

In Neoplatonism, souls are immortal but not completely distinct. They are linked into a world soul that is in contact with both the material and immaterial worlds. This concept of the world soul may be thought of as an early representation of the contemporary Gaia hypothesis for a planetary mind. It also may represent early thoughts on the collective hive minds of bees and other social insects.

Kabbalah

The origins of the Kabbalah are lost in the mists of time. Although some students of Kabbalistic thought claim that it originated in Bronze Age Israel or earlier, it emerges historically in the 12th or 13th century in France or Spain. Kabbalah incorporates aspects of Hermetic thought and Neoplatonism and, although it is not itself a religion, it forms the basis of mystical Jewish religious interpretation. The Kabbalah is no longer a strictly Jewish mode of thought and has also been adopted by adherents of numerous modern occult organizations.

As is true of many occult disciplines, it is not easy for a noninitiate to penetrate the symbolism of the Kabbalah. But some aspects of Kabbalistic thought can be described and interpreted.

There does seem to be a separation between the divine and material realms. At least three versions of Kabbalistic thought have been recognized to deal with this division: the Theosophical approach seeks to describe and understand the divine realm; students of the Ecstatic Kabbalah seek to merge with the divine; and those interested in the white-magic aspects of Kabbalah seek to manipulate the divine. Perhaps because of this third, magical approach, there have sometimes been religious bans on the study of Kabbalistic texts. Other religious authorities have suggested that Kabbalah should be taught to every adult and child.

A basic precept of Kabbalah is its meditation on the nature of God. In Kabbalah the Creator of the Universe seems to have two aspects. One is concealed from mortal knowledge and transcendent to the material realm. The other aspect is revealed through creation and its relation to humanity. It is this second aspect that sustains humans through divinely revealed knowledge. Kabbalists apparently believe that these two aspects are not contradictory, instead they complement one another.

Creation is therefore not completely separate from the Creator. Instead, a series of emanations from the divine sustain the material Universe. These can be studied and understood using both rational and unconscious tools—discourse and symbols. Light is assumed to flow downward from creation and the female aspect of creation is equal to the male aspect, in contrast to the assumption of male superiority in many mainstream religions.

There are 10 emanations or sephirot from the divine that sustain the Universe. These are arranged in a Tree of Life (Figure 4.1) that refers to a tree in the Biblical Garden of Eden. The sephirot are arranged with the highest levels corresponding to the upper branches in the tree's canopy, the middle level corresponding to the trunk, and the lower level corresponding to the tree's roots. In some interpretations, the arrangement of sephirot in the Tree of Life corresponds to three levels of the human soul: with the roots corresponding to instincts, the trunk corresponding to moral virtues, and the higher soul possessing an awareness of the divine. The orbs of the naked-eye planets, the Moon, and the Sun in the Ptolemaic geocentric Solar System model correspond to the various sephirot.

Starting from the highest levels, the sephirot are

Keter: super-conscious will
Chochmah: highest thought potential
Binah: understanding of Chochmah
Daat: knowledge and intellect
Chesad: moral justification for loving kindness (altruism)

Gevirah:	moral justification for judicial severity and strength
Rachamim:	mercy
Yesod:	foundation (strength)
Malkuth:	kingdom (stability)

According to the Tree of Life, evil enters the Cosmos through an imbalance in Gevirah. Although Kabbalah is certainly an esoteric discipline with connection to the Tarot and similar methods of divination, its proposition that lower

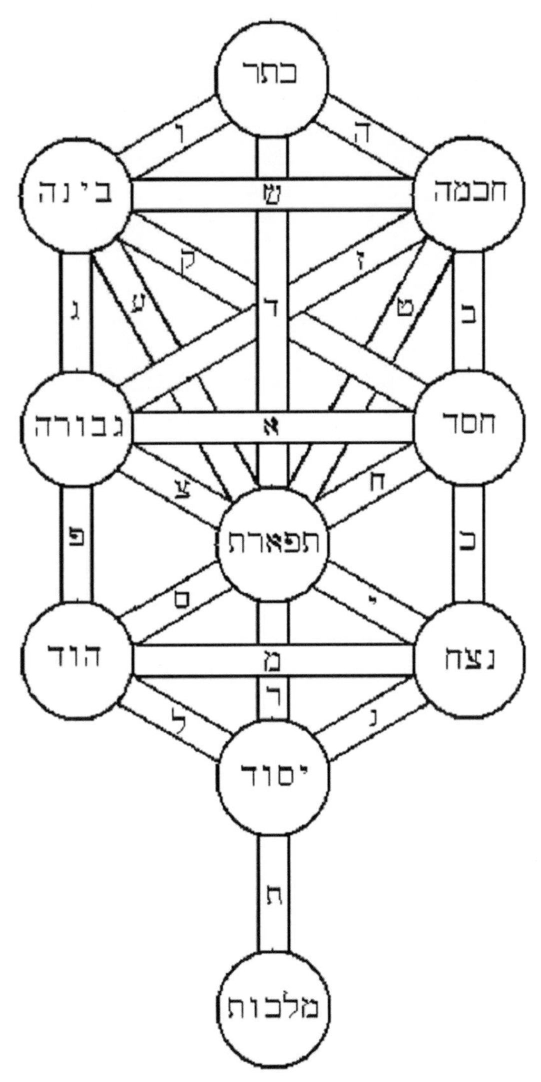

Figure 4.1. The Bahir Hebrew Tree of Life (public domain image uploaded to Wikipedia on January 13, 2006 by Puck Smith).

creation reflects divinity and gave cosmic significance to the daily life of humans.

The Tree of Life and similar constructions do not stand up to rational analysis in today's scientific age. It is perhaps best viewed as a nonrational meditation tool similar to a Tibetan mantra or Japanese koan.

Although it is shrouded in obscurity and open to interpretation, Kabbalah does seem to be consistent with an interaction between the dynamic vacuum and the material world, which is a central scientific concept of this book. The identification of Sun, Moon, and naked-eye planets with divine emanations is a predecessor of the concept of some form of stellar consciousness.

Vedanta

Although it emerged in the early Middle Ages, Vedanta has roots in ancient Indian theological thought. It is considered today as one of the six philosophical schools of orthodox Hinduism. Vedanta is concerned with the relationship between Brahman (the underlying nonmaterial reality) and the material world. Brahman is the supreme cause of material existence; it pervades existence and is eternal. In one of the three versions of Vedanta, Brahman and the world are identical. In a second, they are different but similar. The world is part of Brahman in the third.

The Vedantic assumption that the agency of universal creation is embedded in the material Universe, and is not distinct from it, is similar to Pantheism. A 17th century Western adherent of Pantheism is the Jewish Dutch philosopher Baruch Spinoza. As is true of Vedanta, Pantheism has ancient roots, probably emerging first in the propositions of the 7th century BC Ionian Greek philosopher Thales of Miletus.

CONCLUSIONS

There are major differences between the ancient/medieval and modern human minds. Until the development of scientific method, there were only two forms of knowledge. Revealed Knowledge was obtained directly from a deity (or deities) and is exemplified in Bronze and Iron Age sources such as the Bible and the epics of Homer. The application of Deductive Knowledge, which seems to appear in around the 8th century BC, resulted in the development of superior reasoning techniques. Human civilization could not have developed and advanced without these two formidable roots.

However, at least since the 15th century, the pace of human progress has been guided by the development of scientific knowledge. Science utilizes intuition, which likely is a form of Revealed Knowledge and builds upon this with the development of logical, rational hypothesis. In order for hypotheses to develop into accepted theories, confirmation by inductive experiments and observations is a requirement.

Such free thinking, observation, and experimentation could not have occurred within the intellectual confines of the medieval world. One might imagine that the Middle Ages are in fact bounded by two martyred thinkers: Hypatia, who was flayed in 415 BC by a Christian Fundamentalist mob for her defense of the "pagan" writings in the Royal Library of Alexandria and Giordano Bruno who was burned by Roman Catholic authorities in Rome in 1600 for his writings regarding life on other planets. Galileo narrowly escaped such a fate in 1616 when he was censored by the Church for his published research on the Copernican Heliocentric model of the Solar System. Not to be outdone by Gentiles, Jewish authorities got into the persecution racket when they excommunicated Spinoza in 1656.

The first part of this book can rightly be viewed as a pre-scientific introduction to a topic that may be emerging from the realm of myth, esoteric knowledge, and fiction into the realm of speculative science. However, as will be seen in Part II, ancient and philosophical-theological thought regarding Pantheism heavily influenced the people who developed relativity and quantum mechanics. Our attention will now turn to scientific aspects of the stellar consciousness hypothesis.

FURTHER READING

One of the best descriptions of Hypatia's murder and its aftermath is C. Sagan, *Cosmos* (Random House, NY, 1980). Bruno's contributions and execution are described in the same volume.

An elegant and readable discussion of the contribution to the preservation of ancient knowledge by early medieval Irish monks is T. Cahill, *How the Irish Saved Civilization* (Random House, NY, 1994).

My source for the discussion of Hermetic knowledge and philosophy is *http://wikipedia.org/wiki/Hermeticism* (accessed on May 23, 2014). On the same date, I accessed *http://wikipedia.org/wiki/Neoplatonism* for my consideration of Neoplatonism.

Because of its wide interest in occult circles, there are many sources describing aspects of the Kabbalah. A broad, historical treatment of this mystical discipline can be found online at *http://en.wikipedia.org/wiki/Kabbalah* (Figure 4.1 is from this source). A reprint of a mystical 1935 treatment by an initiate of the Hermetic Order of the Golden Dawn is D. Fortune, *The Mystical Qabalah* (Samuel Weiser, York Beach, ME, 1997).

The online source I used for the discussion of Vedanta is *http://en.wikipedia.org/wiki/Vedanta* (accessed on May 26, 2014).

I also used an online source from the Stanford School of Philosophy to research Spinoza: *http://plato.stanford.edu/entries/spinoza/* (also accessed on May 26, 2014).

Since Galileo has now been "forgiven" by Church authorities, it has recently been proposed that Rabbinical authorities should reopen their case against

Spinoza. Steven Nadler, a professor of philosophy at University of Wisconsin-Madison, discussed this in a May 25, 2014 "Opinionator" piece in the online edition of the *New York Times*. See *http://opinionator.blogs.nytimes.com/2014/05/25/judging-spinoza/*

PART II

Of neurons, tubules, and molecules

Science confronts the "hard question": What is consciousness?

CHAPTER 5

Can we pin consciousness down?

All space, all time,
(The stars, the terrible perturbations of the suns,
Swelling, collapsing, ending, serving their longer, shorter use,)
Fill'd with eidolons only.

The noiseless myriads,
The infinite oceans where the rivers empty,
The separate countless free identities, like eyesight,
The true realities, eidolons.

Walt Whitman, *Eidolons*, from *Leaves of Grass*

In this chapter, I will take on the challenging (if not impossible) task of attempting to pin down a definition for consciousness. Many students of the subject would define this as an attempt to solve the "hard problem". Consciousness is the most basic thing in the world—all of experience, all our internal thoughts and emotions relate to it. Probably, more has been written about consciousness than any other subject. And yet, like Walt Whitman's *Eidolons* (a Greek word defining "spirit images"), consciousness is a phantom concept. As soon as you think you have defined it, it playfully slips from your grasp.

So, rather than attempting to come up with a new, all-encompassing definition, I will review what various other authors have said about it in the past. Consciousness researchers have come from many disciplines: philosophy, theology, psychology, mythology, neurology, and physics. Since I am a physicist and this book is concerned with the possibilities of stellar and astrophysical consciousness, please be tolerant of my emphasis on physicists who have investigated this topic.

First on our list, however, is a generalist. As well as contributing to physics, this early scientist was also a mathematician and philosopher. If you have studied mathematics in secondary school, you have met this fellow.

RENÉ DESCARTES (1596–1650)

In physics, Descartes investigated the rainbow and contributed an early version of the nebula hypothesis, which in modern form describes how the stars and planets condensed from primeval cosmic gas/dust clouds or nebula.

Among his mathematical contributions are Cartesian coordinates, the graphical method taught in secondary school of plotting dependent variables against independent variables.

He is credited with formulating the mind–body problem in philosophy and metaphysics. How does the immaterial mind (or soul) interact with matter in the brain? He also attempted to prove the existence of God by postulating that thought is the essence of mind and matter is the extension of mind. This separation of mind and matter, which may have been related in the inventor's mind to the separation of independent and dependent variables, has come to be known as "Cartesian Dualism." Although dualism in many circles remains a valid metaphysical or philosophical stance, its scientific proponents must deal with several problems. How does the immaterial soul interact with the material substance of the brain? What is more, how did the myriad souls occupy themselves during the 4.5-billion-year evolution of Earth's biosphere to the point at which a sufficiently advanced brain had evolved?

Descartes' essential contribution to consciousness studies was the realization that our consciousness is the only thing of which we can be absolutely certain. His maxim: "I think, therefore, I am" has even been the basis of Hollywood blockbusters such as *The Matrix*. Even if all our experiences are projections or artificial creations of an external agency, we cannot deny that we are experiencing them.

SOME ATTRIBUTES OF A MINDED INDIVIDUAL

Descartes did not attempt to define what a "mind" is. Probably, he took it for granted that most readers would identify it with the Judaic-Christian-Moslem concept of an immortal human soul. But other consciousness researchers during the last four centuries have attempted to define what qualities a human, extraterrestrial, biological, or robotic individual should demonstrate to be considered fully conscious.

The problem is far from intuitive. When I meet a person on the street, I recognize my equal as a conscious being. The same is true when I observe a small cetacean in an aquarium. I am ready to concede that the chimps and elephants I see in the zoo are nearly as conscious as I am. The same goes for parrots and domestic cats and dogs.

About cephalopods I am not so sure. Yes, biologists inform me of the intelligence of squids and octopi but eye contact seems to mean less (to me, at least) with a being so alien.

Furthermore, how far does mind go? Are social insects such as bees and ants minded? How about plants, bacteria, and rocks?

One person who has attempted to deal with this issue in a concise yet authoritative manner is Robert Hanna, a philosopher affiliated with the University of Colorado in Boulder. Hanna suspects that the first question a minded individual asks is "What am I?" The second question is "Who am I?"

According to Hanna, any minded individual will demonstrate four properties—

Consciousness:	the capacity for subjective experiences.
Intentionality:	the ability to direct thoughts toward external things, facts, or beings as well as oneself.
Caring:	the capacity for emotion, desire of others, altruism.
Rationality:	the capacity for logical thought.

I will add another property that seems to emerge at lower, pre-human levels of consciousness. This is volition—the ability to make a conscious choice and act upon it.

Many philosophers and others would of course add other attributes to this list. But, the goal of this book is not to review the long history of speculation on the phenomenon of consciousness; instead, it is to demonstrate that at least some attributes of consciousness are demonstrated by stars and perhaps by the Universe. In preparation for this, attention will now turn to the various definitions of consciousness and speculations regarding its origin by modern researchers.

MURRAY GELL-MANN DISCUSSES REDUCTION

In a reaction to dualism and its religious origin, 20th century scholars developed the concept that "soul", "spirit", or "mind" are different words for the collective product of the functioning of cells in highly developed brains. Proponents of artificial intelligence (AI) consider a modification of this stance in which consciousness is the result of "software" that runs on the "hardware" of the brain.

Nobel Prize–winning elementary particle physicist Murray Gell-Mann describes and defends a version of this stance called "reductionism" in a book that commemorates the centenary of Ramon y Cajal's demonstration of the significance of the neuron in brain functions. According to this metaphysical approach, consciousness emerges as an epiphenomenon (or secondary phenomenon) from brain function when the neural network becomes sufficiently complex.

Reductionism assumes that there is no need for an immaterial soul or spirit and that all aspects of consciousness will ultimately be explained without recourse to non-natural effects by the application of physical laws. Unless you believe in a religion professing a doctrine of the Immortal Soul, the reader should have no problem with reductionism.

However, the correct application of reductionism must apply *all* the physical laws appropriate to the molecular domain of brain function. The proponents of reductionism who attempt to explain consciousness using electromagnetism, molecules, and chemistry only are in error since, as will be discussed in Chapters 6 and 7 in particular, fluctuations in the universal vacuum or quantum foam are also significant at the molecular level. Moreover, according to the well-validated Big Bang theory, a stabilized and greatly expanded vacuum fluctuation is responsible for our Universe. The quantum foam must be treated as the origin of creation, whether you believe in a conscious Creator or a cosmic accident.

Those who neglect these effects are not serving the interests of science. They are certainly not trying dispassionately to solve the hard problem of consciousness. Instead, they are guilty of supporting a dogma in the same manner as the religious person who insists on the existence of an immortal soul and the occurrence of redemption.

IS THERE A UNIVERSAL UNCONSCIOUS?

In the early decades of the 20th century, Sigmund Freud pioneered the new discipline of psychology by separating "higher" conscious brain activities (such as rational thought) from automatic unconscious activities (such as the mechanism involved in bipedal walking). Although Freud believed that all aspects of thought were localized to the physical structure of the brain, not all of his contemporary psychologists agreed.

Notable in his disagreement regarding locality was Carl Jung, who studied the mythologies of many cultures and civilizations widely separated in time and space. Noting many similarities among the mythological records of these diverse peoples, Jung concluded that diffusion was not a sufficient explanation. He postulated that a portion of the unconscious was nonlocal. Spiritually minded people from many origins could tap into this universal unconscious to create their mythologies.

As a practicing psychotherapist, Jung became interested in synchronistic and coincidental events. The most famous of his experiences in this field involved a patient who experienced recurrent dreams of a beetle-like insect. During therapy, as the patient was describing the creature, Jung heard a scratching at the window. When he opened the window, an insect identical to the dream scarab entered.

Interested in the possible scientific significance of such events, Jung enlisted the assistance of a most accomplished (and patient) physicist, Wolfgang Pauli. The collaboration resulted in a joint essay on the topic of synchronicity. Although reductionists may not agree with the results of this effort—especially the pair's collaboration on alchemy, astrology, numerology, and UFOs—it is hard to fault Pauli on intellectual grounds. The Pauli exclusion principle of electron energy levels, derived using the fundamentals of quantum mechanics, has become a necessary tool in physical chemistry.

Jung, with Pauli's assistance, may have been the first to seriously consider the possibility of a universal field that could be tapped in myth. More recent mythologists, notably Joseph Campbell and Jean Houston, have continued these studies and speculations.

The above research efforts are mostly theoretical or deductive. Pauli was most likely attempting to broaden the reach of scientific inquiry by delving into fields far from the mainstream. To advance to the ranks of established scientific theory, the concept of the universal unconscious must be tested experimentally. The British researcher Rupert Sheldrake has taken a step in this direction in his experimental studies of "morphic fields" (nonlocal bonds connecting members of social groups). It is hoped that attempts to replicate the results of his experiments with dogs and other animals will be undertaken.

ERWIN SCHRÖDINGER AND UNIVERSAL CONSCIOUSNESS

Erwin Schrödinger (1887–1961), perhaps most famous for his feline thought experiment, is one of the founders of quantum mechanics. It would certainly be more difficult to theoretically explain wave–particle duality at the base of the quantum world without the Schrödinger wave equation. Like Murray Gell-Mann, Schrödinger was awarded a Nobel Prize in physics. But unlike Gell-Mann, Schrödinger did not favor reductionism.

In 1936, Schrödinger delivered a Tarner lecture at Trinity College, Cambridge, U.K. entitled "Mind and Matter." In this talk, he presented his ideas regarding the material processes associated with consciousness. He felt it unlikely that such a mechanism as complexity could allow consciousness to arise from unconscious material. To replay his metaphor, consciousness is the tutor of the unconscious world.

Schrödinger leans heavily upon the pantheism of Spinoza which describes any inanimate object as a mental construct of God. His thinking is also influenced by ancient sources including the Vedanta.

One of his powerful arguments regards plural minds in the same brain. If mind or spirit is merely the result of neuronal activity and neurons are present in lowly animals such as earthworms, why don't humans (except in pathological cases) have many subminds instead of a single mind?

He discusses the so far fruitless search of science and philosophy for meaning in the space-time construct of the Universe. Many of his examples are drawn from astronomy, biology, and physics. To Schrödinger, God is to be found instead in the realm of spirit—not space-time.

Schrödinger apparently believes that the human mind is a relatively new arrival on the terrestrial evolutionary stage, but mind in general is far more ancient in other life forms. Many of the apparent paradoxes he raises can be rectified by the assumption that consciousness is unitary: there is only one universal mind and organisms that have evolved the appropriate equipment can tap into this omnipresent entity.

As we will see, Schrödinger is not unique in his ruminations on consciousness among those who have delved deeply into the quantum world. One may disagree with the conclusions of these thinkers, but it is difficult to simply brush them aside. Perhaps this is why a 1997 poll revealed that about 40% of working biologists and physicists have strong spiritual beliefs rather than subscribing to strict reductionism.

DAVID BOHM: CONSCIOUSNESS AS PROCESS AND HOLOGRAM

Another physicist who has contributed to consciousness studies is David Bohm (1917–1992). Born and educated in the U.S., Bohm contributed to the Manhattan Project during World War 2, although his application for a security clearance was denied. He collaborated with Einstein before his 1949 harassment by the House Un-American Activities Committee. Ultimately, he left the U.S. and worked in Brazil, Israel, and the U.K. As well as studying mind and consciousness, Bohm contributed to the theoretical development of quantum theory and relativity.

Bohm viewed reality as being composed of a number of levels. Beneath the familiar level of particles, atoms, objects, things, and organisms was an underlying layer of activity. We can grasp certain aspects of this underlying world (which he called the "holomovement") such as photons, electron, or sounds, but this underlying layer is an unbroken totality that cannot be separated into components, defined, or measured.

Aware of the fact that memories are not lost after injuries to the brain, Bohm and others conceived the idea that consciousness operates in a similar manner to a hologram. In other words, brain activities cannot be localized to one region of the brain. The apparent similarity between holograms and brain function is, according to Bohm, in accordance with quantum mechanics and wave theory. He questioned common conceptions of free will, arguing that perhaps thoughts control our actions rather than us controlling our thoughts.

Another very significant contribution was his observation that, in terms of information management, matter acts like a rudimentary or primitive brain. We will apply and enlarge upon this concept in Chapters 6, 7, and 14.

In Bohm's view, it is inappropriate to view consciousness as a "thing." A more accurate representation of consciousness is to consider it as a "process".

Probably, Bohm's major contributions to consciousness studies transcend the realm of theory. His scholarly treatments of the concept opened the door to a later generation of scientists who could investigate this elusive concept from a nonreductionist point of view without endangering or destroying their scholarly reputations.

EVAN HARRIS WALKER CONFRONTS THE HARD QUESTION

Evan Harris Walker, who received his Ph.D. in physics from the University of Maryland in 1964, is perhaps best known for his theoretical work on the quantum theory of consciousness, which is discussed in Chapter 6. In 1999, he authored a popular book on the subject in which he presents his thoughts on the nature of mind and consciousness.

He begins his study with a short description of where we are now. Physics has essentially removed the concept of a personal god from our consideration of reality but, paradoxically, it has shown that the role of the conscious observer is central to all events in the Universe.

From there, he moves on to state that whatever consciousness is, it cannot be defined. Human consciousness, and presumably that of other minds, must be viewed as a totality of experience. Our nervous systems function to take in information from the environment via our senses. Our brains process these data, which are integrated with our ever-changing mental state.

In his discussion, he refutes Descartes by stating that consciousness is not the same as thought. Nor is it perception of the external world through the senses. Instead, consciousness in its pure form is the direct experience of ideas, words, and sensations. It is the carrier of conscious thought but is still present when meditation or deep sleep still the mind.

One analogy he uses is that of a television screen. The television set is unconscious. It is the conscious activity of the viewer that gives meaning to the image on the screen.

Because consciousness interprets data from the far reaches of space, it should not be considered as localized. The space of the entire Universe is spread out before the mind.

Even though Walker refutes Descartes, he expresses the belief that something akin to dualism may be necessary to confront the reality of consciousness. Material phenomena are on one side and consciousness is on the other. Although we cannot define consciousness and must admit its nonphysicality, we can study the consequences of its interactions with the physical world.

Other scholars have echoed and expanded upon some of the concepts discussed above. Two of these are the Austrian astrophysicist Erich Jantsch and the American quantum physicist Henry P. Stapp.

FURTHER READING

My reference for René Descartes was the online Stanford Encyclopedia of Philosophy (*http://plato.stanford.edu/entries/descartes/*). For an essay on minded individuals, see R. Hanna, "What is the self," *Perspectives on the Self: Conversations on Identity and Consciousness, Annals of the New York Academy of Sciences*, **1234**, 121–123 (2011).

For an essay on reduction, see M. Gell-Mann, "Consciousness, reduction, and emergence," *Cajal and Consciousness, Annals of the New York Academy of Sciences* (ed. P. C. Marjian), **929**, 41–49 (2001).

The epic tale of the collaboration and friendship of Carl Jung and Wolfgang Pauli is beautifully presented in A. I. Miller, *Deciphering the Cosmic Number: The Strange Friendship of Wolfgang Pauli and Carl Jung* (Norton, NY, 2009). Two excellent studies of myth are J. Campbell, *The Inner Reaches of Outer Space: Metaphor as Myth and as Religion* (Harper & Row, NY, 1986) and J. Houston, *A Mythic Life* (HarperCollins, NY, 1996).

Rupert Sheldrake presents some of his results in a well-referenced volume, R. Sheldrake, *Dogs that Know when Their Owners are Coming Home and Other Unexplained Powers of Animals* (Three Rivers Press, NY, 1999).

Erwin Schrödinger's thoughts on consciousness and related material have been collected, reprinted and published in E. Schrödinger, *What Is Life?: With Mind and Matter and Autobiographical Sketches* (Cambridge University Press, Cambridge, U.K., 1967). The 1997 poll on religious and spiritual beliefs of working biologists and physicists is discussed by G. Easterbrook, "Science and God: A warming trend?" *Science*, **277**, 890–893 (1997).

The most popular and familiar of David Bohm's contributions to the quantum theory of consciousness is D. Bohm, *Wholeness and the Implicate Order* (Routledge & Kegan Paul, London, 1980). An online source describing the life and contributions of this theoretical physicist is *http://en.wikipedia.org/wiki/David_Bohm* in which many of his publications are cited.

The concept of Universe as hologram has been further developed and advanced by others. Notable contributions to this field have been made by Stanford neurophysiologist Karl Pribham, who has applied this concept to brain function. Much of this work is reviewed and cited in a popular treatment by M. Talbot, *The Holographic Universe* (Harper Perennial, NY, 1992).

Evan Harris Walker reviews his quantum theory of consciousness and attempts to pin down this elusive concept in the very readable book E. H. Walker, *The Physics of Consciousness* (Perseus Books, Cambridge, MA, 1999). In this volume, he also describes the events that led him to concentrate his research on the nature of mind.

You can access some of Erich Jantsch's thoughts by checking out E. Jantsch, *The Self-Organizing Universe: Scientific and Human Implications of the Emerging Paradigm of Evolution* (Pergamon Press, NY, 1980). Henry R. Stapp presents some of his contributions in the monograph H. R. Stapp, *Mind, Matter, and Quantum Mechanics* (Springer Verlag, NY, 1993).

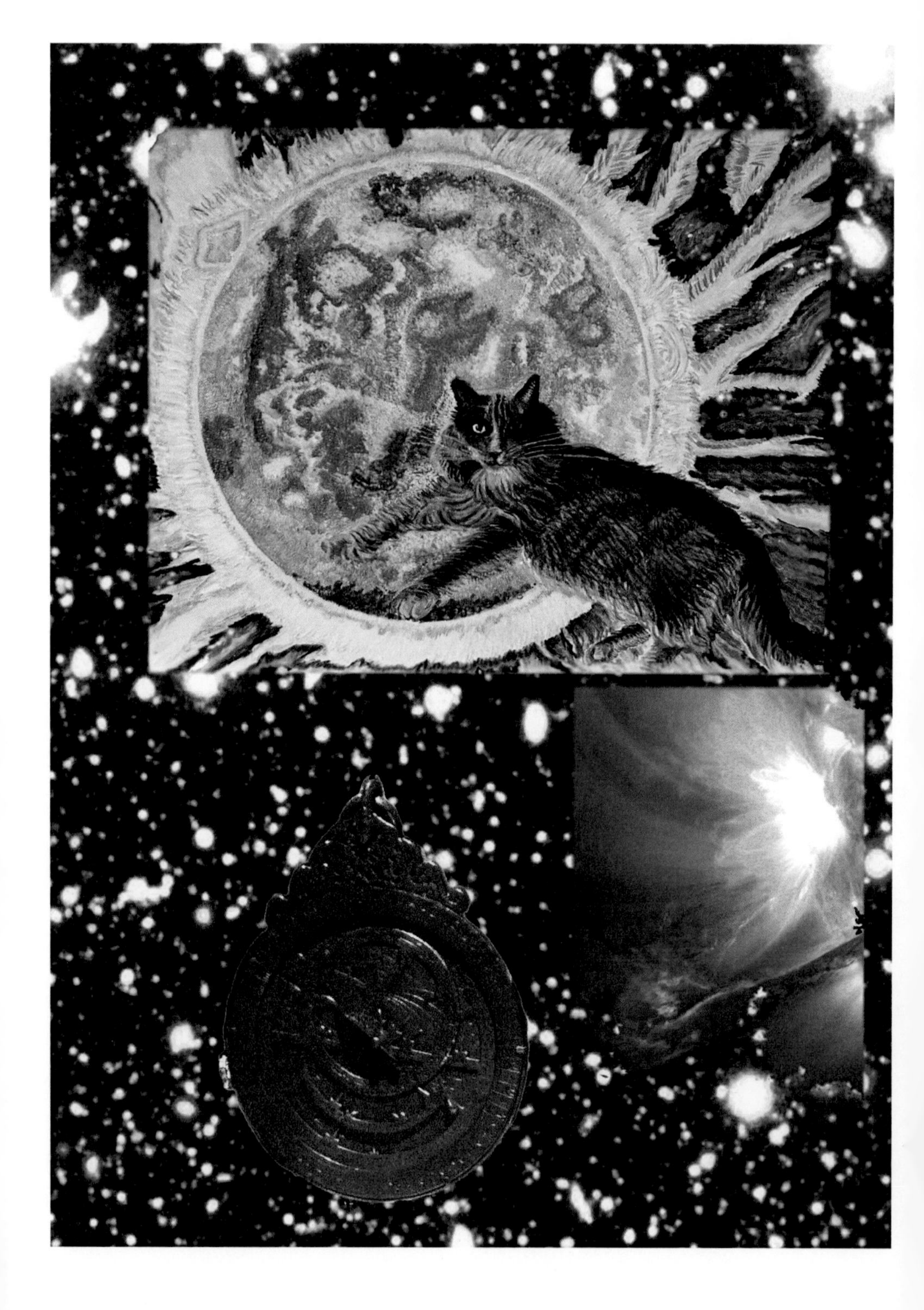

CHAPTER 6

Consciousness and quantum mechanics

We are the music-makers,
And we are the dreamers of dreams,
Wandering by lone sea-breakers,
And sitting by desolate streams;
World losers and world forsakers,
On whom the pale moon gleams;
Yet we are the movers and shakers
Of the world for ever, it seems.

Arthur O'Shaughnessy, *Ode*

In the early 20th century, we lost forever the comfortable, mechanistic world of Isaac Newton. The clockwork models of classical mechanics, which had dethroned Earth from a central position in the Universe, were themselves replaced by the strange, counterintuitive theories of relativity and quantum mechanics.

When we examine the very fast or massive, classical approximations must be replaced by special and general relativity. Moreover, when we examine the molecular, atomic or subatomic worlds, nondeterministic quantum theory holds sway. Only at intermediate scales do classical calculations still apply to a close approximation.

If you feel uncomfortable about this, you are not alone. Relativity theory, after all, brought us the very mixed blessing of nuclear energy. But if we reject the quantum, we must also give up computers, cellphones, GPS. In fact, much of our contemporary civilization depends heavily on the applications of quantum theory.

PARADOXES AND COMPLEMENTARITIES

Quantum physics begins at the dawn of the 20th century. To explain a discrepancy between observed ultraviolet electromagnetic emissions and observational data, the German physicist Max Planck (1858–1947) proposed a

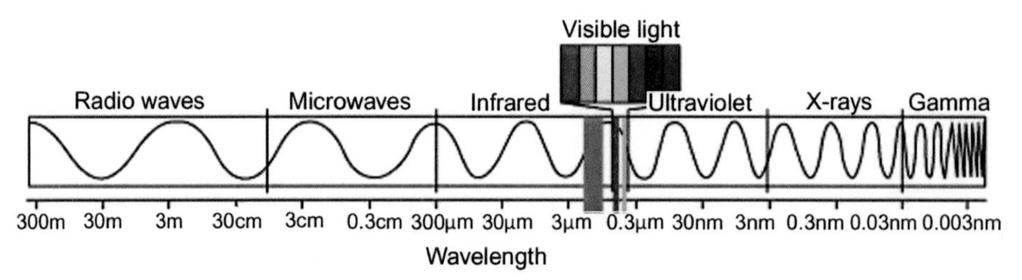

Figure 6.1. The electromagnetic spectrum (courtesy NASA).

radically new model for light. The wave and particle characteristics of light were composed in a "quantum", the smallest unit of electromagnetic radiation.

The energy of a quantum of light, called the photon, would be directly proportional to the frequency and inversely proportional to the wavelength. Short waves, such as gamma rays and X rays, are more energetic, penetrating and particle like. Longer wavelength radio waves are less energetic and wave like (Figure 6.1). In the visual spectrum, blue light photons are more energetic than red photons.

Studying the photoelectric effect, in which high-energy photons can eject electrons from their atoms, Albert Einstein demonstrated that, although they are without mass, photons have linear momentum. This photon linear momentum (which in classical physics is the product of a particle's mass and velocity) is called radiation pressure. In the Sun (and other stars), radiation pressure from photons generated in the Sun's deep interior balances the self-gravitation of the Sun's enormous mass. This balance prevents our star from collapsing. Radiation pressure has also been applied in recent years to the solar sail, an in-space propulsion technique that requires no fuel, just the pressure of impinging solar photons.

It is not too difficult mathematically to consider a photon as a dual entity, something that can act both as a wave or particle, depending upon which characteristic the experimenter is investigating. But the concept is certainly counterintuitive. How do we derive a mental picture of an object that is sometimes a particle (a dimensionless point with no extension in space) and sometimes a wave (something with peaks and troughs that is not confined to a single point)? This is especially true for particles with mass, such as electrons. These were found to have wave characteristics, as do photons. The wavelength of such "matter waves" decreases with particle mass.

One person who accepted this challenge was Louis de Broglie, a French physicist working in the U.K. In the early 1920s, de Broglie and others developed the "wavicle" model. Imagine that you can greatly magnify the beam from your flashlight or laser pointer and that you can slow the photons from their enormous velocity (300,000 kilometers per second or 186,300 miles per second in vacuum) to a standstill. When you zoom in upon the individual photons in the beam, you don't "see" tiny points. Instead, you see blobs like tiny pieces of Jell-O or clay. Somewhere within each of those blobs is the probable location of a photon, but it is wrong to take such mental models too literally. To paraphrase Niels Bohr, we can derive abstract expressions for quantum events, but we cannot actually grasp the nature of this realm.

One person who contributed mightily to the mathematical formalism of quantum mechanics is the German physicist Erwin Schrödinger. His wave equation, which works for particle quanta moving far less than the speed of light in vacuum, applied probability theory to matter waves. His wavefunction should be viewed not as a representation of the subject particle but instead as the probability of the particle being there.

According to another quantum pioneer, Pascual Jordan, it is the act of observation that collapses this probability wave to a measurable result. To the great consternation of many classical physicists, mind had re-entered science!

One idea that led to further development of quantum physics was complementarity. The wave and particle aspects of a quantum are contradictory, but both are necessary for a complete description of the object. Only one of the two quantities in a complementary pair can be measured at a time.

Another German, Werner Heisenberg, proposed the famous uncertainty principle. Consider, for example, two aspects of a moving quantum: velocity and position. As we try to pin one down with greater accuracy, the other becomes more uncertain. It was later demonstrated that the uncertainty principle can be derived from the Schrödinger wave equation.

Heisenberg turned his attention to a classical experiment: the double-slit experiment (Figure 6.2). Suppose that you direct a monochromatic light (or monoenergetic electron) beam toward an opaque screen with two closely spaced slits of the same approximate size as the wavelength of the impinging quanta in the beam. An interference pattern will be produced on a screen placed on the side of the slits opposite the beam.

This interference pattern was a well-known demonstration of the wave nature of light. But Heisenberg (and others) wondered what would happen if you cut the beam intensity down until you had only one quantum. Intuition would say that the quantum would pass through one slit or the other and that

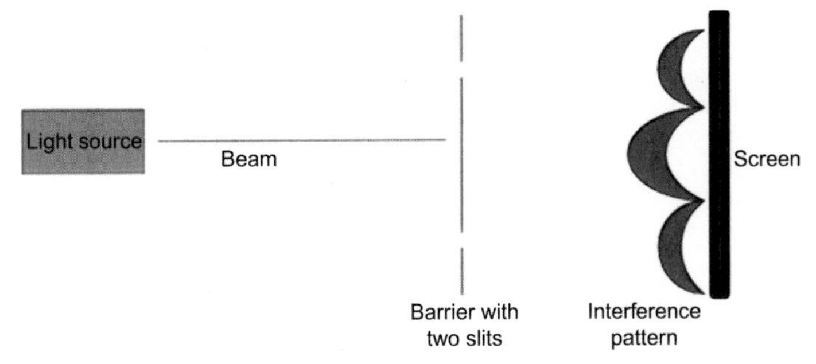

Figure 6.2. Two-slit photon interference pattern experiment schematic.

no interference pattern would be produced. But Heisenberg's analysis had the opposite conclusion: the probability wave function associated with the single quantum would simultaneously pass through both slits. An interference pattern would be produced as the single quantum interfered with itself!

When the experiment was performed and the results replicated, the conclusion was unmistakable. An interference pattern was formed when a single quantum was beamed toward the two slits (Figure 6.2). In 1927, Bohr and Heisenberg met in Copenhagen to try to arrive at an interpretation of this mysterious new quantum realm.

The "Copenhagen interpretation" that emerged from this meeting became the mainstream philosophical approach to quantum mechanics for decades. In this operational approach, it is tacitly assumed that there is in actuality no quantum world. It is not the role of physics to describe what nature is; physics instead describes what we can say about nature.

This approach allowed a generation of physicists to apply the techniques and results of quantum mechanics to electronics, chemistry, and the realm of the atomic nucleus without worrying about deep philosophical implications. But as quantum thought evolved and applications began to enter the mid-level domain we inhabit, many scholars became uncomfortable with the Copenhagen Interpretation.

One demonstration of quantum effects in the mid-level world is Schrödinger's famous cat. In this thought (or *gedanken*) experiment, Puss is equipped with water, food, a litter box, a scratching post, and appropriate toys to divert her during the experiment. She is locked within an opaque container for an hour. The container is equipped with a vial containing a deadly poison that is opened only if a radioactive substance decays. The half-life of this substance is the duration of the experiment. At the conclusion of the experiment, there is a 50% chance the cat is alive and a 50% chance that she is dead. But since radioactive decay is a quantum phenomenon, the situation inside the box is more complex: until observed, the cat's wave function has not collapsed and she is suspended in an intermediate, double state. One way out of this paradox is to accept the cat as a conscious observer who has witnessed the result of the experiment and thereby caused her wave function to collapse to either alive or dead.

Alternative interpretations to Copenhagen were introduced and evaluated as the 20th century advanced. Some of these, which are discussed in more detail by Bruce Rosenblum and Fred Kuttner, are introduced below.

Extreme Copenhagen

According to Bohr's son Aage and Ole Ulfbeck, the Copenhagen interpretation is incomplete. It should be expanded to claim that the micro-world is nonexistent. In this approach, objects such as atoms are nonexistent. Although this approach evades the role of consciousness, it is hard to take it seriously in a world where atoms have been imaged and the energy locked within the atomic nucleus has produced copious quantities of electricity, controlled cancers, and destroyed two cities killing as many as 200,000 people.

Decoherence and consistent histories

In an attempt to eliminate the role of the observer, some physicists have considered the case of an atom whose wave function is simultaneously in two separate boxes. A photon is allowed to pass through each box. It only interacts with the "actual" atom. After it bounces off the particle, its wave function interacts with the wave functions of the atoms in the detection equipment. Averaging over all these wave functions yields a classical-type probability for the existence of the atom in one of the two boxes. But even the developers of this decoherence scheme concede that a model of consciousness is required to explain how the experimenter is aware of the atom's location.

A development of this concept, the consistent histories approach, postulates an atom moving through the early Universe when organic observers had not yet evolved. As the atom interacts with other particles in a probabilistic fashion, a series of possible histories develops. A system that gathers and utilizes information must develop to pin down the atom's true history. Consciousness and free will seem to be inevitable in this interpretation.

Many worlds

Developed by Hugh Everett in 1957 during his graduate student days, the many worlds interpretation is perhaps the most famous and most discomforting of the alternatives to Copenhagen. To illustrate how the many worlds interpretation avoids the collapse of the wave function by consciousness, consider the case of poor Puss once again. Instead of her suspension in a dead/alive confusion, the Universe splits. In one Universe, she purrs and meows. In the other, she does not. Of course, there is no mechanism proposed for this infinitely splitting Cosmos. Furthermore, although consciousness is not required to collapse the wave function, this concept of avoidance increases the power of consciousness in the Universe. If every conscious decision splits the Universe, consciousness is very powerful indeed! Another philosophical issue with the many worlds interpretation is raised by Rosenblum and Kuttner. If there is a near-infinite number of alternate universes, does the word "reality" have any meaning?

Transactional

This interpretation invokes symmetry in time: the wave function can move both forward and backward along the time axis. To paraphrase John Cramer's remarks quoted by Rosenblum and Cramer, as we observe the light from a star 10 light years away, retarded waves from the star have been traveling in our direction for 10 years. The advanced waves generated by the interaction of the photon with our eyes have reached 10 years in the past. The interaction between the photon and the conscious observer is still very much in the picture.

Bohm's hidden variables

In Bohm's attempt to derive a deterministic interpretation of quantum mechanics, he invoked a "quantum potential" that acts upon all particles and insures that, on average, the results demanded by Schrödinger's equation are produced. This "hidden variable" connects observer and observed. Although Bohm's approach seems to be in conflict with the very well validated theory of special relativity, it does seem to allow an undivided Universe. In Bohm's view, his interpretation does not negate the requirement for consciousness.

Mermin's "Ithaca interpretation"

David Mermin, a physicist associated with Cornell University in Ithaca, New York has attempted to reinterpret quantum mechanics by defining two types of probability. Classical probability is a subjective concept: it is a measure of the observer's ignorance. Quantum probability, on the other hand, is objective and the same for all observers. According to the Ithaca interpretation, quantum mechanics tells us what correlates with physical reality and what does not. This approach assigns consciousness to a real realm that transcends today's concepts of physical reality.

Quantum information

In this approach, the wave function is identified neither with the physical system under study or a description of that system. Instead, it represents information regarding the possible measurements that can be performed on that system. Essentially, therefore, quantum mechanics is only about consciousness, not about the physical world.

Other interpretations

The above sections discussed the major interpretations of quantum mechanics that have been proposed as alternatives to the Copenhagen interpretation. Other interpretations consider changes in the rules of logic or invoke physical phenomena that have never been observed, according to Rosenblum and Kuttner. However, it seems clear from the above discussion that most interpretations accept or require a role for consciousness.

QUANTUM THEORIES OF CONSCIOUSNESS

As physicists developed a deeper understanding of the quantum world and the interaction between observer and observed, some began to develop theories regarding the nature of consciousness and its interaction with matter. Would Descartes agree with this boldness, or is he doing cartwheels in his grave?

Henry Stapp's grand superposition

Stapp, an American physicist affiliated with the University of California at Berkeley, has developed a quantum theory of consciousness that maintains the concept that quantum wave functions collapse from probable to real during their interaction with consciousness. Unlike the theories to be presented below, he believes that the collapse is global rather than confined to various structures or regions within the brain. In Stapp's view, there are essentially two realities: physical and mental. Wave functions collapse when the conscious mind selects one among the many possibilities. Although Stapp's theory (like all the others discussed below) is controversial, he does present supporting evidence from observations of brain function.

Evan Harris Walker: Synaptic clefts and quantum tunneling

Evan Harris Walker, another American physicist, has also developed a theory of the interaction between consciousness and the physical brain. Unlike Stapp, he has located regions in the brain which seem to be supportive of quantum events. The region he selects is the synaptic cleft. Nerve cells in animal brains branch in a tree-like manner. Synapses are structures in neurons that allow electrical or chemical signals to be transferred from one of the myriad neurons in a brain to a neighboring neuron. The cleft, or spacing, between synapses from two adjacent neurons is typically 20 nanometers (nm).

Walker calculated that this synaptic cleft could function as an electrical potential well because of this tiny spacing and the electrical charge on adjacent synapses. He then considered what could happen to a charged particle (such as an electron) within this potential well.

In the macroscopic world, you might throw a rubber ball against a wall trillions of times with perfect confidence that the ball will repeatedly bounce off the wall. In the quantum world, things are a bit more complex. At the quantum level, whenever the trapped particle encounters the barrier, there is a finite possibility that it could instantaneously tunnel through and reappear on the other side. At least one Nobel Prize has been awarded for work on this effect, and quantum tunneling has found wide application in electronics, nuclear physics, and related fields.

As was the case for Stapp, Walker correlated the predictions of his theory against the observations of brain functions. But unlike Stapp's theory, Walker's theory considers particular locations in the brain's neuronal network and a well-studied and documented quantum event. However, as discussed in Chapter 7, Walker's theory had unexpected ramifications. If there was a finite possibility of information or thought tunneling from one brain location to another, might it also manifest in someone else's brain? For the first time, studies of telepathy and other paranormal concepts became respectable.

Microtubules and quantum correlation

Perhaps the major significance of Walker's theory is that it opened the door to other researchers. Are there other structures than neurons in the brain that could host quantum events? What is more, were there alternative quantum processes to tunneling?

One of the foremost microbiologists of the late 20th and early 21st centuries, Lynn Margolis of the University of Massachusetts, considered this problem. She arrived at the conclusion that 25–50 nm structures present in all living cells, called microtubules, could play a quantum role equivalent to that of neurons. Since all cells possess these structures, all cells may be conscious in a sense. Typical neurons in animal brains contain millions of microtubules, which are identified as neurotubules.

The distinguished mathematical physicist, Roger Penrose of Oxford University, has collaborated with the American anesthesiologist Stuart Hameroff to develop a comprehensive theory of quantum consciousness centered upon neurotubules and quantum nonlocality (also called quantum coherence or quantum entanglement).

To gain an appreciation of quantum coherence, consider the following example. A particle (P) and its antiparticle (A) are produced from a high-energy gamma ray. One of the quantum properties—let's call it "spin"—is conserved. Since the gamma ray has zero spin, if A has an up spin, P must have a down spin and vice versa.

But we cannot pin down which has spin up and which has spin down until we perform an experiment and collapse the wave function through observation. If we observe antiparticle A in our experiment and determine that its spin is up, then the spin of particle B must be down. Even if the particle and antiparticle are separated by light years after they are generated, this spin correlation must occur instantaneously.

This was very upsetting to Albert Einstein since his well-verified theory of general relativity postulates that the absolute speed limit for matter or information transfer in our four-dimensional space-time is the speed of light. He conducted a three-decade dialog with Niels Bohr on this concept and presented something like the above example in the Einstein–Podolsky–Rosen (EPR) thought experiment to support his disbelief in "spooky action at a distance" and his conviction that something was amiss with quantum theory. In 1964, John Bell derived a rigorous mathematical approach that led a team directed by French physicist Alain Aspect to perform the EPR experiment. Sadly for Einstein and happily for Bohr, quantum nonlocality was confirmed. Since

Aspect's initial work was published in 1981, many other experiments have confirmed his results.

Penrose and Hameroff denote their model using the acronym Orch-OR (for orchestrated objective reduction). One of the provisions of this model is that consciousness is noncomputable. In the view of Pennrose and Hameroff, future computer systems will not become conscious, no matter how advanced or complex they become.

Consider, for example, Schrödinger's cat. Poor Puss is somewhat confused, being simultaneously alive and dead before being observed. Penrose and Hameroff propose a process called "objective reduction" (OR), which causes the unfortunate beast's wave function to collapse to either alive or dead, even without an observer. OR is a physical process that is beyond the bounds of current quantum theory. The operation of OR is based upon quantum gravity, a theory that Penrose has contributed to and, in its ultimate development, will expand our understanding of black holes and the Big Bang.

According to this theory, OR occurs because of a significant difference in the space-time geometries (and gravitational effects) between the two states. Two states in microtubules within neurons could support OR effects on timescales necessary for neural functions. Quantum entanglement between adjacent neurons or quantum tunneling between adjacent neurons would also play a role. The OR effect in the brain collapses the wave functions of observed objects and things associated with them.

OR events can be viewed as elements of a proto-consciousness field that is tied in with space-time geometry. A tiny element of experience is essentially a moment of proto-consciousness. Another way to look at OR is to consider it as a bridge between the quantum and classical worlds.

The OR theory is, of course, very controversial. As Hameroff and Penrose discuss in their recent review paper, some initial criticism disagreed with the premise that microtubules at body temperature could sustain quantum events. Recent experiments by a Japanese team have demonstrated, however, that such events do indeed occur.

Penrose and Hameroff note that one of the effects of Alzheimer's disease is death of neuronal microtubules. It is not impossible that future verification of aspects of the OR theory will lead to clinical applications. Nevertheless, of more relevance to the main topic of this book is the fact that the OR theory offers the possibility of consciousness in at least some stellar environments: neutron stars. This happy congruence will be further discussed in Chapter 23, in particular.

Bernard Haisch and zero-point fields

The reader may rightly wonder if there are candidate fields for proto-consciousness. He or she may also realize that, although animals have neurons and all living cells have microtubules, it is very unlikely that such structures will be found in stellar interiors. Fluctuations in the zero-point field may be a candidate for proto-consciousness. As well as affecting the behavior of neurons and microtubules, such fluctuations contribute to intramolecular forces. If these conjectures are correct, any star cool enough to have molecules in its upper layers may have a form of consciousness.

In 1928, the British physicist Paul Dirac compared the vacuum with a sea. If a particle, say an electron, is manifested, this can be treated as the crest of an ocean wave. The corresponding trough of that wave constitutes the (then-hypothetical) antiparticle of the electron, the positron. When the positron was discovered by Carl Anderson of the California Institute of Technology in 1932, studies of such vacuum fluctuations became a subdiscipline of theoretical physics.

It is now accepted that at tiny physical scales and in very small time intervals, the vacuum is a dynamic medium with particle–antiparticle pairs constantly springing in and out of existence. Tremendous energies exist with negative and positive energies canceling each other so that the net result is zero. The minimum permissible energy of a quantum mechanical system is called "zero-point energy." About 13.7 billion years ago, something happened to stabilize a vacuum fluctuation. From a volume less than a cubic millimeter, all the matter and energy in our Universe, in fact the geometry of space-time, sprang. The Big Bang originated from such a stabilized vacuum fluctuation.

In 1948 the Dutch physicist H. B. G. Casimir was studying the forces between and within molecules. He noted that classical electromagnetic theory could not completely account for the way these van der Waals forces vary with particle separation. He concluded that the cause for this discrepancy might be vacuum fluctuations and he designed an experiment to test his conjecture. In 1996, S. K. Lamoreaux at the University of Washington conducted the first experimental verification of Casimir's concept. Many others have since replicated these results.

The so-called Casimir effect works as follows. Consider an experimental arrangement consisting of two electrically conducting plates and associated equipment. The plate separation is of the order of a micron or less. At such small separations, not all vacuum fluctuations can fit between the plates. Therefore, there is a vacuum pressure that contributes to the force between the plates.

The American physicist Bernard Haisch, among others, has recognized the possible connection between vacuum fluctuations and consciousness. To encapsulate his view, the zero-point field is essentially identical to the Penrose/Hameroff concept of the proto-consciousness field. This is essentially a pantheistic view. All matter at the molecular level or higher would be imbued with at least a low level of consciousness. It is significant that in this approach, proto-consciousness derives from the basic creative agency of the Universe: the dynamic vacuum. Whether you subscribe to a single creation in a Universe optimized for intelligence or a multiverse with ours just happening to be just right, a proto-conscious vacuum foam can fit your worldview.

Interestingly, the optimum particle separation or size to manifest a conscious Casimir effect allows this effect in the neuronal gap and the microtubule. During the 1980s, when U.S. governmental agencies began to investigate the (thus far very elusive) possibility of applying the Casimir effect to space propulsion or terrestrial energy production, the American space scientist Robert Forward derived the optimum plate separation in his conceptual Casimir machine to be about 20 nm.

The concept of vacuum fluctuations contributing to consciousness and the organization of the Universe has captured the imaginations of other scientists. See, for example, the short paper referenced below by R. R. Laughlin of Stanford University's Physics Department.

CONCLUSIONS

As the above discussion outlines, one can make a very compelling case for quantum effects playing a role in the phenomenon of consciousness. It is reasonable to study consciousness in humans before we can generalize to other life forms or stars. However, science requires that mathematical conjectures be tested by experiment or observation before they can be accepted as confirmed theories. As Chapter 7 reviews, the process of conducting consciousness experiments on human subjects has serious pitfalls.

FURTHER READING

There are many excellent sources describing the history of quantum mechanics, how quantum mechanics relates to nature, and the role of the conscious observer in quantum calculations. In the preparation of this chapter, I have used two of them. One, co-authored by a physicist and classical scholar (both affiliated with George Mason University in Fairfax, VA), is M. Kafatos and R. Nadeau, *The Conscious Universe: Part and Whole in Modern Physical Theory* (Springer-Verlag, NY, 1990). Also used in this chapter is a popular volume co-authored by two physicists at the University of California, Santa Cruz, B. Rosenblum and F. Kuttner, *Quantum Enigma: Physics Encounters Consciousness* (Oxford University Press, Oxford, U.K., 2006). Either of these sources offer a far more comprehensive introduction to the confrontation between modern physics and consciousness than I can produce in one short chapter.

Henry Stapp's theory is discussed online (*http://en.wikipedia.org/wiki/Henry_Stapp*). It is also outlined in the above-cited reference by Rosenblum and Kuttner. Stapp has also described his approach and its correlation with psychological data in a number of technical papers and books including H. P. Stapp, *Mind, Matter, and Quantum Mechanics* (Springer-Verlag, NY, 1993).

For a popular treatment of Evan Harris Walker's quantum consciousness theory, check out his book E. H. Walker, *The Physics of Consciousness* (Perseus Books, Cambridge, MA, 2000). A more technical treatment is E. H.

Walker, "The nature of consciousness," *Mathematical Biosciences*, **7**, 131–178 (1970). Although Walker's book treats the subject, there is also a good online source for information on quantum tunneling (*http://en.wikipedia.org/wiki/Quantum_tunnelling*).

Some of the contributions of Lynn Margolis to consciousness studies are described in L. Margolis, "The conscious cell," *Cajal and Consciousness, Annals of the New York Academy of Sciences* (ed. P. C. Marjian), **929** (2001).

Roger Penrose's early thoughts on consciousness are expressed in R. Penrose, *The Emperor's New Mind: Concerning Computers, Minds, and the Laws of Physics* (Oxford University Press, Oxford, U.K., 1989). His more developed concepts and his collaboration with Stuart Hameroff are discussed in the above-cited book by Rosenblum and Kuttner. See also S. Hameroff, "Consciousness, the brain, and spacetime geometry," *Cajal and Consciousness, Annals of the New York Academy of Sciences* (ed. P. C. Marjian), **929** (2001) and R. Penrose, "Consciousness, the brain, and spacetime geometry: An addendum. Some new developments on the Orch OR model for consciousness," *Cajal and Consciousness, Annals of the New York Academy of Sciences* (ed. P. C. Marjian), **929** (2001).

The most recent review paper devoted to the OR theory is S. Hameroff and R. Penrose, "Consciousness in the Universe: A review of the 'Orch OR' theory," *Physics of Life Reviews*, **11**, 39–78 (2014). Two papers cited in this paper describing experimental work on quantum effects in microtubules are S. Sahu, S. Ghosh, K. Hirata, D. Fujita, and A. Bundyopadhyay, "Multi-level memory-switching properties of a single brain microtubule," *Applied Physics Letters*, **102**, 123701 (2013), *http://dx.doi.org/10.1063/1.4793995* and S. Sahu, S. Ghosh, B. Ghosh, K. Aswami, K. Hirata, D. Fujita, and A. Bandyopadhyay, "Atomic water channel controlling remarkable properties of a single brain microtubule: Correlating single protein to supramolecular assembly," *Biosensors and Bioelectronics*, **47**, 141–148 (2013).

Some of Penrose's earlier thoughts on the subject are in R. Penrose, "Quantum computation, entanglement and state reduction," *Philosophical Transactions Royal Society London A*, **356**, 1927–1939 (1998).

One nontechnical source describing the dynamic vacuum that underlies existence is H. Genz, *Nothingness: The Science of Empty Space* (Perseus Books, Reading, MA, 1998). For a discussion of the possible significance of vacuum fluctuations in consciousness studies, see B. Haisch, *The God Theory: Universes, Zero-Point Fields, and What's Behind It All* (Weiser Books, San Franscisco, CA, 2006).

Robert Forward contributed widely to U.S.-government funded studies of unconventional space propulsion techniques and energy sources. His calculations on energy from vacuum fluctuations are included in the report R. L. Forward, "Alternate Propulsion Energy Sources," AFRPL TR-83-067, Air Force Rocket Propulsion Laboratory, Air Force Space Technology Center, Space Division, Air Force Systems Command, Edwards Air Force Base, CA 93523 (December, 1983). Forward subsequently published some of his results in the open literature. See R. L. Forward, "Extracting electrical energy from the vacuum by cohesion of charged foliated conductors," *Physical Review B*, **30**, 1700–1702 (1984). This paper is available online at *http://www.zpower.com/wap/documents/ZPEPaper_ExtractingElectricalEnergyFromTheVacuumByCohesionOfChargedFoliatedConductors.pdf*

The American physicist R. R. Laughlin has published some speculations on the role of vacuum fluctuations in the Universe's organization. See R. R. Laughlin, "The cup of the hand," *Science*, **303**, 1475–1477 (2004).

CHAPTER 7

The Uri Geller affair

Hey nonny no!
Men are fools that wish to die!
Is't not fine to dance and sing
When the bells of death do ring?
Is't not fine to swim in wine, And turn upon the toe
And sing hey nonny no!
When the winds blow and the seas flow?
Hey nonny no!

Anon. (Elizabethan), *Hey Nonny No!*

Most scientists work hard to insure that their experiments are as objective as possible. But when you work with human subjects in the real world, strange and unexpected things may happen. One reason for this is the human desire to "dance and sing" and to "swim in wine" rather than mud. It must be realized that in competitive, capitalist society, people will be tempted by possibilities of self-advancement, even those people who have been acclaimed as talented psychics.

The late 1960s and early 1970s were a unique time in the United States, perhaps in most of the Western world. After the near-catastrophe of the Cuban Missile Crisis in 1962, the Cold War had been frozen into a ritualistic mold. While it had perhaps become less likely that global nuclear war would be conducted by the two main players, there was a constant jockeying for position and a constant effort to curry favor in the developing world. Since direct military engagement between the U.S. and U.S.S.R. seemed to be remote, the contending powers had found a new arena: outer space. With America's victory in the Moon Race, one might have assumed that some form of U.S. Protestantism might become the dominant world faith. But, instead, photos from deep space contrasting verdant Earth with the barren Moon gave rise to the environmental movement and a rebirth of the ancient Earth goddess, Gaia.

Proxy warfare was also conducted by both powers. In Vietnam, the U.S. was involved in the hopeless defense of a corrupt dictatorship against the forces of a disciplined, nationalistic communist army. Because the intellectual competition of the Cold War had given rise to a vast, young and energetic intelligentsia in the U.S. and conscription continued to send young cannon fodder to Vietnam, the stage was set for a student rebellion.

As part of this youth rebellion, an experimental counterculture equipped with radical new art forms developed. As well as enjoying more sexual freedom than their parents, the young hippies had access to a large variety of psychotropic drugs. Watered-down versions of quantum consciousness theories appeared in popular publications.

Geographically, the burgeoning U.S. counterculture of the late 1960s and early 1970s was centered on the east and west coasts of North America. On the shores of the Pacific, the city of San Francisco became the hippie Mecca. One educational institution in Palo Alto (California), Stanford University, was destined to play a major role in the attempts to validate and apply quantum consciousness. And, paradoxically, the effort would be funded not by a rock star or counterculture guru, but by that mainstay of Cold War competition: the Central Intelligence Agency (CIA).

THE QUANTUM SPIES MEET THE HIPPIE PHYSICISTS

In the popular imagination, intelligence agents lead a glamorous, albeit short life. They drive fast vehicles, have liaisons in exotic locales with attractive (and dangerous) sexual partners, engage in shoot-outs with ruthless bad guys, and trade huge sums of money for world-saving information. Much of this is a Hollywood creation. The average spy is a bureaucratic civil servant who hopes to live long enough and quietly enough to enjoy his or her retirement pension.

Rather than risking life and limb in the manner of the mythical Agent 007, a typical intelligence agent might like to rise in the morning, drive or take public transit to work, engage in a calm day's work, and return home in the evening. Wouldn't it be nice if, rather than flitting off to Istanbul, one could simply sit at one's desk, use an appropriate drug or meditation technique, and get inside the mind of one's opponent on the far side of the Iron Curtain.

Someone in the home offices of the Central Intelligence Agency must have read a preliminary version of Evan Harris Walker's paper (cited in the previous chapter) in *Mathematical Biosciences*. Using a well-known, validated and applied quantum phenomenon called "quantum tunneling," it might be possible to do just that.

Much of the work would be done at the Stanford Research Institute (SRI), a spin-off of Stanford University. In the U.S. intellectual climate of the late 1960s and early 1970s, Stanford was not the only institution of higher learning to separate its defense-funded projects from its academic pursuits.

Instead of conducting preliminary tests on members of the general public, much of the research was conducted on individuals with alleged psychic abilities. Although this factor of the SRI-centered research effort can perhaps be faulted, it is certainly understandable. Some might argue that the best way to proceed would have been to use screening tests on volunteers from the student population and study those subjects with the highest scores in greater depth. But most research grants are limited in scope and funding. Grant recipients are thereby encouraged to utilize limited funds in the most efficient way possible to obtain results, and insure that the grant is renewed or expanded in the next fiscal cycle.

As well as psychologists, physicists (mainly from northern California institutions of higher learning) became involved in the SRI effort to evaluate the science behind paranormal effects. Some of the young physicists who were attracted to the effort were part of the Fundamental Fysiks Group based in Berkeley (California).

In retrospect, it is very understandable to me why so many talented, young scientists would be attracted to this study, which was destined to be viewed by their more conventional colleagues as a fringe effort.

As I experienced on the opposite side of North America, academic and industrial opportunities for young physicists essentially vanished in late 1969 and early 1970. America had proven its technological prowess to the world by beating the U.S.S.R. to the Moon. After the flights of Apollo 11 and Apollo 12, funding sources began to very rapidly dry up. Lay-offs were common in industrial research laboratories (as I sadly learned and experienced myself) and many academic programs were curtailed or eliminated. So, many young physicists began to participate in the counterculture. In some cases, at least, they began to reject the Copenhagen interpretation and investigate the new (and very fashionable) alternative interpretations of quantum mechanics, some of which were introduced in Chapters 6 and 7.

PSYCHIC OR MAGICIAN?

One of the alleged psychics who visited SRI was an Israeli named Uri Geller. Although the funding agencies would have been more interested in psychic powers with direct application to intelligence gathering, such as telepathy (or mind reading), some of the researchers conducting the screening tests had a more general interest in alleged psychic abilities.

Geller's performance on double-blind screening tests was exceptional. For instance, he was able to reproduce with remarkable accuracy drawings that had been previously sealed in an envelope and successfully guess eight times in a row the value of a die within a metal box. He presented to SRI staffers a demonstration of another alleged psychic power—telekinesis or psychokinesis (PK)—the ability to manipulate objects at a distance by the force of will alone. He demonstrated this ability by bending keys and kitchen cutlery. The physicists in the audience were fascinated; many wondered how he could transfer the required energy from one location to another by the agency of mind.

To discuss and further investigate paranormal spoon bending and related phenomena, the scientists associated with SRI and the Fundamental Fysiks Group participated in a number of conferences. These, almost from the start, were destined to increase the controversy swirling around this infant field of research.

For example, one such conference was conducted in Reykjavik (Iceland) in November 1977. The proceedings of this conference (dubbed *The Iceland Papers*) were edited by Andrija Puharich, a friend and supporter of Geller. The foreword to this book, however, was authored by Brian D. Josephson of Cavendish Laboratory, Cambridge, U.K., a Nobel Prize winner in physics.

Two papers in *The Iceland Papers* deal with the phenomenon of paranormal metal bending. One, by Another British scientist, John B. Hasted of Birkbeck College, University of London, considers experimental tests on various subjects attempting to bend paperclips and similar objects at a distance. Hasted was later criticized for poor experiment design, which to some readers invalidated his positive results.

Evan Harris Walker and Richard D. Mattuck of the University of Copenhagen, on the other hand, collaborated in an attempt to apply quantum theory to the problem. In their analysis, they address the issue of where the required energies of 0.01–100 Joules could come from. They calculated that energy conservation was no problem: in a PK event, consciousness could make use of the energy present in matter from random quantum fluctuations. However, they admitted that the second law of thermodynamics—that entropy (or disorder) increases in any isolated system—was violated.

In the analysis of how minded stars might alter their galactic velocities, to be presented in Chapter 15, I will discuss stellar PK as one of the options affecting this trajectory alteration. In an active star there is considerable energy to spare.

However, as the SRI effort progressed, the difficulty with performing experimental investigations of the paranormal with human subjects became more apparent. Uri Geller developed into an international celebrity, authoring best-selling books and receiving very respectable lecture fees. At one of his lectures, the magician James "The Amazing" Randi was in the audience. Randi later discussed cutlery bending with the lecture organizer and reported that the alleged PK episode could be duplicated in a magic trick.

This is the only scientific debate in which I have been privileged to know well meaning and honest people on both sides. In my discussions with Evan Harris Walker, I learned that it is very unlikely that Geller could have cheated on the videorecorded screening studies at SRI. Two other well-known physicists have echoed Walker's opinion: Hal Puthoff of the Institute for Advanced Studies in Austin Texas and Edgar Mitchell, who founded the Institute of Noetic Sciences after visiting the Moon aboard Apollo 14.

On the other hand, I was introduced at a cocktail party in around 2000 to a retired editor who worked for Time Warner. Although I have forgotten the gentleman's name, I very well remember the incident. After learning that I was a physicist who had some acquaintance with PK but was neutral regarding its validity, he walked over to the bar and picked up a fork.

As we talked, he claimed to have coordinated the lecture that led to Randi's claim that Geller is a magician. But as I watched, the fork began to bend!

I was amazed. He laughed when I asked: "Where does the energy come from?" He told me that this was a trick taught to him by Randi. He had promised The Amazing Randi that he would never reveal the exact nature of this sleight-of-hand demonstration. He also told me that my question was typical of physicists, who are apparently quite gullible as a class when confronted by the artistry of a master magician.

SEEKING A MIDDLE PATH

This retold history of the Uri Geller affair presents an open-minded, albeit skeptical scientist such as myself with a conundrum. On one side of the issue, I have never witnessed what I would classify a genuine PK event. My only direct experience with the subject was an invitation to observe metal bending as a magic trick. On the other side of the issue, there is the testimony of the "expert witnesses" described above, the fact that the phenomenon is apparently allowed within the confines of quantum theory, and a whole lot of anecdotes available online and in popular books.

Some reject the phenomenon out of hand and others accept it as true in spite of the fact that Geller's popular performances may have utilized sleight of hand. This difference in opinion may have more to do with an individual's personal metaphysics than the PK phenomenon.

One person who has striven for a middle path between these extremes is Jack Sarfatti, an original member of the Fundamental Fysiks Group. When I met Sarfatti at a recent conference dealing with interstellar travel propulsion physics, I learned that he currently believes that we should allow the results of the original screening tests on Geller into our considerations because they would be very difficult, if not impossible, to duplicate by sleight of hand. On the other hand, Geller's more popular demonstrations of metal bending before audiences may be tainted as data because of his abilities as a master magician.

For my own part, I have tried to develop and maintain a related viewpoint. Perhaps Geller did display genuine psychic talent in the screening tests at SRI. Perhaps he was thereafter tempted by the vast sum of money he could earn by supplementing this talent with his magic skills while performing before large paying audiences. I suspect that a substantial fraction of the readership, as well as myself, might succumb to such a temptation if the potential reward were substantial.

In my more negative moments, however, I sometimes deviate from the middle path. Is it possible that the CIA has engineered a vast conspiracy theory, similar to the one subscribed to by some advocates of UFOs? Maybe "they" have developed a method of psychic mind control and manipulation of electronic equipment at a distance. If the U.S. is ever invaded by a technologically equal foe, or if the powers that be decide to end democracy for good, these tools will be carted out from Area 51 (or some other secret depository) and used for good or evil! All of the talk about Uri Geller's authenticity, sleight of hand, and the trick I saw performed by the retired editor are part of the conspirators' cover!

SHOULD WE REOPEN THE CASE ON PK?

As mentioned in the Introduction, one place where I presented the case for minded stars was Paul Gilster's *Centauri Dreams* blog. One of the respondents to my piece, obviously acquainted with physical theory and technique, suggested that the case for a weak PK force be reopened in an original way. He/she suggested that rather than attempting to repeat the screening tests of the SRI group, modern researchers might try to measure attempts to manipulate a remarkable state of matter called a "Bose–Einstein condensate."

In Chapter 6 the consciousness theory of Roger Penrose was discussed. Penrose and his collaborator Stuart Hameroff have suggested that one type of star, a neutron star, might have a core composed of this material and that a Bose–Einstein condensate could support consciousness.

A Bose–Einstein condensate is an unusual state of matter that is composed of bosons (particles with integer spin). Explored theoretically in the 1920s by Satyendra Bose and Albert Einstein, the first successful generation of a Bose–Einstein condensate was by Eric Cornell and Carl Weiman of the University of Colorado in 1995. They received the 2001 Nobel Prize in physics for this achievement.

What is special about this state of matter is that particles in it collapse into their lowest accessible quantum state. Particles in a Bose–Einstein condensate behave like a single, joined super atom. Essentially, quantum properties usually found at the microscopic, atomic, or molecular level are observed in a macroscopic substance.

Originally, it was necessary to cool the gas atoms in a sample to near absolute zero to observe the properties of a Bose–Einstein condensate. Recently, researchers associated with IBM have been able to create a Bose–Einstein condensate at room temperature. Various applications for this state of matter may now be possible.

One of the interesting properties of this state of matter is superfluidity. Cooled near absolute zero, gases such as helium-4 disobey the normal rules of surface tension. Superfluid samples in the form of a thin film have been observed to actually crawl up the side of its container.

Here, then, is a possible application for consciousness researchers seeking to experimentally investigate the reality or nonreality of PK. They could request their subjects to concentrate on a superfluid placed in another room and viewed through a window. By the power of their will, they would attempt to control the rate at which the superfluid climbs the walls of its container. A well-conceived experimental program could perhaps check for conscious effects in the superfluid and thereby test the Penrose–Hameroff theory. However, it would be a challenge separating the conscious effects produced by the experimenter, the subject, and the superfluid (if any).

FURTHER READING

To my knowledge, the most comprehensive, informative, and entertaining document regarding the subject of this chapter was authored by MIT physics professor D. Kaiser, *How the Hippies Saved Physics* (Norton, NY, 2011).

Many of the physicists who participated in the activities of the Fundamental Fysiks Group and related organizations went on to author successful popular books on aspects of quantum consciousness: G. Zukov, *The Dancing Wu Li Masters: An Overview of the New Physics* (Morrow, NY, 1979); F. A. Wolf, *Star Wave: Mind, Consciousness, and Quantum Mechanics* (Macmillan, NY, 1984); and F. Capra, *The Tau of Physics: An Exploration of the Parallels between Modern Physics and Eastern Mysticism* (Shambhala, Boulder, CO, 1975).

The two works on paranormal metal bending discussed above are J. B. Hasted, "Paranormal metal-bending," *The Iceland Papers*, pp. 95–110 (ed. A. Puharich, Essentia Research Associates, Amherst, WI, 1979) and R. D. Mattuck and E. H. Walker, "The action of consciousness on matter: A quantum mechanical theory of psychokinesis," *The Iceland Papers*, pp. 111–160 (ed. A. Puharich, Essentia Research Associates, Amherst, WI, 1979). Check out *en.wikipedia.org/wiki/John_Hasted* for an online source discussing criticism of Hasted's experimental technique.

Edgar Mitchell has documented his experiences during the Apollo 14 lunar mission and his subsequent investigations into aspects of consciousness. You can read his take on the Uri Geller affair in E. Mitchell and D. Williams, *The Way of the Explorer: An Apollo Astronaut's Journey through the Material and Mystical Worlds* (Putnam, NY, 1996).

The speculations of Roger Penrose and Stuart Hameroff regarding the ability of a Bose–Einstein condensate within a neutron star to support consciousness is available on-line. Check out R. Penrose and S. Hameroff, "Consciousness in the Universe: Neuroscience, quantum space-time geometry and Orch OR theory," *Journal of Cosmology*, **14** (2001), *http://journalofcosmology.com/Consciousness160.html*

An online source describing the Bose-Einstein condensate is A. Z. Jones, "Bose–Einstein Condensate" in *About.com Physics* (*http://physics.about.com/od/glossary/g/boseeinstcond.htm*). For information on the recent production of a Bose–Einstein condensate at room temperature, see D. Johnson, "Bose–Einstein condensate made at room temperature for first time" in the December 10, 2013 online edition of *IEEE Spectrum* (*http://spectrum.ieee.org/nanoclast/semiconductors/materials/bose-einstein-condensate-made-at-room-temperature-for-first-time*). This piece also discusses the possible applications of Bose–Einstein condensates that may be developed. For a discussion of superfluidity, see A. Z. Jones, "Superfluid" in *About.com Physics* (*http://physics.about.com/od/glossary/g/superfluid.htm*).

CHAPTER 8

Gaia and her sisters

All ye woods, and trees, and bowers,
All ye virtues and ye powers
That inhabit in the lakes,
In the pleasant springs or breaks,
Move your feet
  To our sound
Whilst we greet
  All this ground
With his honor and his name
That defends our flocks from blame.

John Fletcher, *To Pan* (from *The Faithful Shepherdess*)

For thousands of years, civilized humans had treated their environment, our planet, almost with disdain. Yes, they could grow crops on its soil, draw water from its lakes, pasture their herds on its prairies, and draw countless fish from its oceans. They treated our world as an infinite source of riches and an infinite sink for untreated sewage and other pollution. After all, didn't many of the holy books state that humans have divinely granted dominion over the Earth and all nonhuman forms who dwell within her?

And then, suddenly, in the late 1960s and early 1970s, this began to change. To many or most people in the developed world, Earth is now recognized as finite. Humans are part of an interlocking web of life, a biosphere, rather than being lords of the planet. Our collective actions, as shepherds of this planet, can determine the future of this ecosystem and our very survival as a species. It harkens back to the classical nature god Pan, and even earlier to the Earth-mother Gaia. But how did this happen? And what does it mean for suprahuman consciousness?

APOLLO AND VOYAGER

As is true of most societal changes, this one was impossible to predict. It begins in that horrible and dramatic year, 1968. The war in Vietnam dragged on.

More and more of the disaffected young drew toward the counterculture. Some of the political and religious leaders who could have perhaps provided salvation from the morass had been assassinated.

The U.S. and U.S.S.R. were locked in a race to send a human expedition to our Moon and return it safely to Earth. Both space programs were recovering from fatal accidents aboard Apollo 1 and Soyuz 1 the year before. In the closed society of communist Russia, space mission planners had a major advantage over their American competitors. Planned space missions in the more open U.S. were publicly announced months in advance. So, Russian astronautics teams could rush their preparations to beat the competition to a space goal by a few days, weeks, or months. Consequently, it seemed to much of the world that Russia was ahead in the space race.

But this was not really the case. The Russian super booster had a nasty habit of exploding early in flight while the U.S. Saturn V was moving smoothly toward operational status. It is probable the Russians had secretly conceded the Moon-landing race to the U.S. in 1968. However, they realized there was a major propaganda advantage in being the first space program to safely send humans to the vicinity of the Moon. So, the Soviet program began to concentrate on a stripped-down version of Soyuz called Zond. Robots and animals were sent around the Moon. A few even survived entry. It seemed to American planners that a human cosmonaut could circle the Moon aboard Zond in the near future. Almost certainly, the Russians would schedule a launch to upstage the American effort.

This led to a Cold War ruse being adopted by the American space agency. It is remarkable that none of the thousands involved in planning and conducting NASA missions spilled the beans to the press.

It was announced that Apollo 8 would be launched in mid-December 1968. This was to be the first all-up test of the Saturn V with a human crew aboard the Apollo capsule. According to the official press release, this would be an Earth-orbital flight.

However, U.S. intelligence agencies knew from their spy satellites that the Soviet Proton rocket used to loft Zond toward the Moon required lengthy preparation. As Apollo's launch date approached, NASA managers called a hasty news conference. They announced that the mission plan had been uprated because of the condition of their "bird" and the capabilities of the crew. Apollo 8 would not simply repeat earlier flights and orbit the Earth, it would orbit the Moon.

Russian space managers tried to quickly prepare a mission to upstage Apollo. But it was a futile effort. With an international audience of a billion or more, Apollo lifted off smoothly, carrying its three-man crew toward their appointment with the Moon and with destiny.

To add to the U.S. propaganda advantage, the U.S. astronauts elected to read responsively from the *Book of Genesis* as their craft circled the Moon. This reading was accompanied by television images showing a verdant Earth rising above a stark lunar horizon.

Figure 8.1. Apollo 8 photo of Earthrise over the lunar horizon (courtesy NASA).

During the lunar orbit phase of Apollo 8, a classic photo of Earthrise was taken by the crew. After debating for decades which of them was responsible for this image, the astronauts finally proclaimed that it was a joint undertaking. One loaded the film; a second pointed the camera; the third snapped the image.

It's likely that mission planners and astronauts hoped for a religious revival based upon the readings and images. But it is doubtful that they suspected this revival to encompass a very ancient religion. The Earthrise photo (Figure 8.1) sparked the environmental movement and the concept that all terrestrial life is linked in an analog to the ancient Earth Mother called the Gaia hypothesis.

More than two decades after the Apollo 8 astronauts splashed down at the end of their mission, another significant photo was snapped from the depths of space. This one was destined not only to show Earth's unique nature in the Solar System as a living world, but also to demonstrate its isolation in space.

In early 1990, NASA's Voyager 1 probe, one of humanity's first emissaries to the Milky Way galaxy, was cruising on the edge of the interstellar void. Mission controllers requested the craft to turn its cameras one more time toward the inner Solar System. One of the images (Figure 8.2) shows the Earth as a pale-blue dot. The arrow denotes the position of the Earth, submerged in a beam of sunlight scattered by the camera lens.

All that we have collectively accomplished or dreamt of, all human lives and deaths, all our civilizations, wars and holocausts are on that one small dot billions of kilometers from the questing cameras aboard a small spacecraft.

Earth is an incredibly precious living jewel. We will have to search far to find its equal as an abode of higher life. We are linked to other life forms in the planet's intricate biosphere. But is this marvelous system in a sense conscious, or may it become so?

A LIVING PLANET

Naturalists, as well as astronomers, were intrigued by the concept of Earth as a living planet. Instead of considering individual life forms and species involved in an endless Darwinian competition sparked by random mutations, environmental changes, mass extinction events, etc., it was necessary to develop a scientific approach to the concept of an Earth-embracing web of life. The American microbiologist Lynn Margulis and the British chemist and inventor James Lovelock collaborated on the first stage of this effort in 1974. In their monumental paper in the journal *Icarus*, they demonstrated how the biosphere and atmosphere of our planet cooperate to form a self-regulating system. The interactions between life and environment and the coevolution of various life forms (such as hummingbirds and orchids) were investigated by naturalists like David Attenborough.

Life does seem to optimize the environment it exists in. Evolution does seem to increase novelty among living species. This self-organizing nature of the biosphere, which has been going on for around 4 billion years, is hard to credit to the operations of blind chance alone.

Figure 8.2. Earth as a pale-blue dot taken by Voyager beyond Neptune (courtesy NASA).

Stages of Gaia

Scientists have a pretty good idea of how life and the environment have coevolved on our planet. It is possible to examine this process by considering its sequential stages.

Shortly after the formation of the Solar System, about 4.7 billion years ago, the Sun turned on. Sunlight and particle streams emanating from the Sun dissipated our planet's primeval hydrogen/helium atmosphere. Impacting comets and outgassing from the planet's interior brought the oceans and a reducing atmosphere. This early atmosphere was rich in methane, ammonia, and carbon dioxide. Earth's environment at this point may have resembled that of Saturn's satellite Titan at the present time.

Energized by lightning and high-energy solar photons, life developed in this hydrocarbon "soup". The first terrestrial microbes were anaerobic, thriving in methane and emitting oxygen as a waste product.

Gradually, as oxygen levels increased, Gaia experienced her first environmental crisis. If the anaerobic microbes had any capacity for reason (which I seriously doubt), it must have seemed that all the sewers in the world had burst at once. They retreated to methane-rich, oxygen-poor environments such as deep-ocean volcanic vents. Their mutated cousins, which were fine about oxygen, began to dominate the planet's surface biosphere.

Some of these aquatic microbes developed the ability to cooperate and form colony organisms such as modern-day jellyfish. By 500 million years ago, some invertebrates such as clams had developed specialized cells that could form protective covers. Others developed crude, internal support systems that evolved into skeletons. Fish, the first vertebrates, had evolved.

Meanwhile, photosynthesis had developed in aquatic plants. More of the atmosphere's carbon dioxide reserve was converted into oxygen. Gradually, a balance was produced as plants and animals coevolved. Animals consume oxygen and emit carbon dioxide as a waste product. Plants are net consumers of carbon dioxide and emitters of oxygen.

As atmospheric oxygen levels increased, ultraviolet sunlight reacted with stratospheric oxygen molecules. The protective stratospheric ozone layer formed. As ultraviolet levels decreased, life began to colonize the land.

Colonization of the land must have been a tedious process. The descendants of the first vertebrates to emerge from the sea, the amphibians, must even today spend part of their life cycles in an aqueous environment.

By 300 million years ago, the first reptiles had evolved. Large land organisms were able to spend their entire lives on land. As the climate warmed, the largest reptiles, the dinosaurs, dominated our planet's surface. Were it not for the random impact of a $\sim$10 km asteroid or comet in what is now the Yucatán Peninsula about 65 million years ago, mammals (such as humans) and small feathered dinosaurs (birds) would not have stood much chance of evolving very far.

Today, Gaia is experiencing another environmental crisis as the human population peaks and people in the developing world expand their consumption levels to meet that of consumers in the developed West. Pollution levels are increasing, global climate change is occurring, and many species are in decline.

It is possible that we are collectively conducting an uncontrolled experiment on Gaia's resiliency. Hopefully, the many feedback loops that constitute this global, interlocking system will be able to adjust to the actions of our species.

I also sometimes wonder about an alternative possibility. What if climate change is a collective intelligence test of the human species? What is more, what happens if we fail?

A CONSCIOUS GAIA?

According to proponents of orthodox Darwinian theory, evolution is a blind, totally random process. As Earth's environment is modified due to oxygen excess, supervolcanoes, or impacting celestial bodies, organisms attempt to adapt to the changes. Some of the offspring of living organisms will be subject to random mutations that can be passed along to the next generation. If a mutant is more suitable to the altered environment, it tends to fill the many ecological niches vacated by those species that could not adapt and have become extinct.

Although this is a powerful theory that explains many aspects of the biosphere, many scientists question certain aspects of it. First, how did such organs as the eye evolve? Vision of one form or another has evolved many times and an eye must be perfect to perform its function. There simply is no halfway house.

Another problem is the coevolution of different species. How, for example, did certain species of hummingbirds develop beaks optimized to pollinate certain species of orchids that have evolved to serve those particular beaks?

Furthermore, there is the problem of hive organisms. Most readers will be familiar with social insects such as bees, ants, and termites. One of these small creatures is rather helpless and ineffectual in the larger scheme of things. But, in many cases organized around the matriarchal queen, they can construct elaborate structures as a team and live according to a rich social code. Even more dramatic is the slime mold amoeba. For much of its existence, this creature exists as a solitary cell. Then, according to some mysterious signal, myriads of these tiny creatures link up to form a slug-like organism that crawls along the forest floor and eventually erects a tower. From the top of this immense construct (from the viewpoint of a single amoeba), reproductive cells can be spread over a wide range.

It is difficult to believe that random evolutionary mutations in response to environmental modifications can produce these effects. One possibility is that random may not be truly random. A simple demonstration that randomness may actually work to create order and complexity is the Game of Life, which was devised by the British mathematician John Horton Conway in 1970. An initial configuration is input to a computer screen with some simple rules. In one version of this game, a cell interacts with its eight neighbors. For instance, a cell might die (by simulated underpopulation) if it has one or zero neighbors, survive if it has two or three neighbors, and die (by simulated overpopulation) if it has more than three neighbors. A dead cell with exactly three living members can return to life in a form of reproduction. Players of this game, which can be downloaded to all or most computers, notice that elaborate structures are built and evolving cell communities can send out explorers to colonize other portions of the screen.

So, the mathematical rules of the Universe may result in more order being produced in a biosphere as time progresses. Another way of looking at this is that Gaia is actually an evolving information field as well as a network of interconnections between organisms and the environment.

After a distinguished career including directing solar observatories in Capri and in the Canary Islands, the Swedish astrophysicist Arne Wyller retired to Santa Fe (New Mexico) where he devoted his creative efforts to the problem of evolution and the biosphere. He concluded that there was simply not enough time for conventional Darwinian theory to result in the rapid rate of evolution that ultimately resulted in humanity.

Wyller concluded that, instead of organic evolution proceeding as a series of random acts, life had coevolved with a planetary mind field. This entity, which might be considered as a nervous system for Gaia, is not to be confused with the creator God of most scriptures and is constrained by the laws of physics. But, operating within these constraints, this minded Gaia directs evolution with a thirst for novelty.

Gaia's most recent evolutionary experiment, according to Wyller, is humanity. The main function of this element of the biosphere, which is far more intellectually and technologically capable than other species, is to reach out into the Universe to search for Gaia's sisters.

The concept of a conscious Gaia may be viewed as parallel to the morphogenetic field of Sheldrake and the universal proto-consciousness concept of Penrose and Hameroff, which were discussed in Chapter 6. Indeed, all of these concepts and nonrandom randomness may be different complementary aspects of the same principle.

Whether or not Wyller is correct in his assessment of Gaia's present-day nervous system may prove to be immaterial since we may today be constructing a planetary mind. A walk down any urban street in the developed world will reveal to the observer a massive change in human consciousness that is currently underway. With cellphones and other mobile electronic devices, more and more people are constantly linked at the speed of light to a worldwide flow of information that we have dubbed the Internet. If Gaia does not have a nervous system at present, she may well have one soon!

THE HUNT FOR GAIA'S SISTERS

Today, the hunt is on for sisters of Gaia: other living worlds in our region of the Milky Way. In our Solar System there is almost certainly no equivalent to Earth. Yes, primitive life may be found on Mars, beneath the frozen seas of Jupiter's satellite Europa, in the hydrocarbon lakes of Saturn's satellite Titan, or in water geysers emitted from Saturn's small satellite Enceladus. If life evolved within the confines of the Solar System's birth nebula rather than on our planet's surface, some speculate that bacteria may be found even in the tails of comets.

What is different, however, is that although life may have achieved a foothold in such environments, it has not modified the local environment to its advantage, as has Gaia. Our beautiful blue planet is a unique gem within the Solar System. In the far future, daughters of Gaia may be transplanted to surfaces or interiors of other Solar System (and galactic) objects. But these will be Gaia's offspring, not her sisters.

A number of astronomical observatories on Earth's surface and in space are devoted to locating extrasolar planets: planets circling other stars. As of the time of writing (June 2014), about 1800 have been discovered.

Three basic techniques are used to search for extrasolar planets. The first to be used (starting in the 1990s) is the study of radial velocity variation of the parent star caused by its much less massive planet. When the planet is between the star and us, the star is pulled slightly in our direction. This gravitational velocity change causes the star's spectral lines to shift slightly to the blue. When the planet is on the far side of its star, the star moves slightly in the direction of the planet and the star's spectral lines are shifted a bit toward the red.

If the star has a significantly large proper motion across the sky, the planet's gravitational tug on the star as it circles the star can cause the star's position to wobble. This astrometric technique only works for low-mass, near stars.

NASA's Kepler Space Observatory has had great success applying the transit technique. If the planet orbits its star in the correct orientation, it will pass directly in front of the star (from our point of view). This will cause the star's apparent brightness to slightly decrease in a periodic fashion.

Finally, if you have a huge telescope at your disposal and very clear skies, you can place an occulting disk in front of the star to greatly reduce its glare. If the star is near enough to our Solar System and the planet's orbital position is not too close to the star, it is just possible using advanced image-processing techniques to image the planet.

The first planets located tended to be hot Jupiters, massive objects very close to their parent stars. Jupiter-like worlds in elliptical orbits and brown dwarfs (objects intermediate in mass between large planet and small stars) were also discovered. As techniques were refined, smaller planets were discovered as well as many multiple-planet solar systems similar to ours. Many planets apparently reside in stable orbits in binary star systems.

Difficult as it is, finding an extrasolar planet is not enough. The next step is to assess its habitability. Such a procedure is based upon a conceptual approach developed in the 1960s by Rand Corporation astrophysicist Stephen Dole and described in the popular press in collaboration with famed science author Isaac Asimov. Following this procedure, the astronomer first notes the luminosity of the host star relative to the Sun. Then, he or she assumes that an exact duplicate of the Earth circles that star in a nearly circular orbit. The life zone or ecosphere is defined by the planet-orbital location close to the star at which all water on the surface boils and the location at which all water freezes. In our Solar System, Mercury and Venus are within the inner boundary of the ecosphere and are considered to be lifeless. Earth is near the center of the ecosphere and is ideally situated for the evolution of life. Mars is on the outer fringe of the ecosphere and is considered to be marginally habitable.

Following Dole's pioneering approach, other astrophysicists expanded the definition of the ecosphere. What happens if a hypothetical planet has a different inclination than Earth's or is in a more elliptical orbit? If a planet has more greenhouse gases (carbon dioxide, water vapor, methane) than Earth, the ecosphere can extend farther out from the Sun than in Dole's initial projections. If it has fewer greenhouse gases in its atmosphere than the Earth, the ecosphere can extend a bit closer to the central star.

Dole considered that stars much hotter, more massive, and bluer than our Sun might not be good candidates for a habitable planet search since these stars do not live long enough for advanced life to evolve. Although stars much redder, cooler, and less massive than the Sun are very common, Dole concluded that such red dwarf stars are less likely than Sun-like stars to have habitable planets because the ecosphere is narrow and so close to the central star that stellar flares might hinder the development of life.

Many now consider Dole's analysis to be too conservative. Initially, the bounds of the ecosphere were determined by the amount of sunlight striking a planet's surface. It is now known that certain life forms live deep underneath Earth's surface deriving energy from geothermal processes. Geothermal, tectonic, and tidal effects on the satellites of Jupiter and Saturn may have allowed life to develop in environments far from the Sun.

Additionally, it has been learned that the low-mass cutoff for stars capable of hosting habitable worlds may have been too conservative. Unexpectedly, extrasolar planets have been detected within the ecospheres of red dwarf stars.

But location may not be enough in determining whether an extrasolar planet hosts a sister of Gaia. To learn what a distant life-bearing world would look like, NASA tried an experiment with our Earth in 1992.

Previously, the U.S. space agency had launched the Galileo probe to Jupiter using a space shuttle. In the wake of the Challenger accident, astronauts were reluctant to fly with the high-energy upper stage designed to boost the probe directly toward that giant planet. As an alternative, they applied a less capable rocket to send the craft on a trajectory that would allow several close passes of inner planets to derive enough gravitational energy for a Jupiter flight.

Figure 8.3. Earth and Moon from Galileo (courtesy NASA).

On one of the passes of Earth, Galileo's instruments were turned toward our planet. Images were snapped of the Earth and Moon, which show in sharp contrast the difference between a life-bearing planet and a dead world (Figure 8.3).

Earth also shows sharp methane and ozone spectral lines and is a very active emitter of (artificial) radio signals. We can search for these signs of life on nearby extrasolar planets in the habitable zone when our equipment becomes sensitive enough.

If we learn that all or most habitable worlds bear Gaia-like biospheres, this will be very exciting. Such a discovery will be an indication that life can arise in any possible location and that self-organization into more complex forms is built in to the evolutionary process.

However, if life-bearing worlds are rare or nonexistent, this sobering discovery will be of value as well. It will indicate that either random effects or divine intervention sparked life on our world and we had better act as stewards of Gaia and carefully protect her while we plan to plant her seeds in nonterrestrial locations.

FURTHER READING

There are many popular books that outline and describe the events of the great Moon race of the 1960s. One of the nicest, in my opinion, is J. Barbour, *Footprints on the Moon* (Associated Press, NY, 1971).

The vastness of space and the tiny expanse of humanity's domain as evidenced by Voyager's "Family Portrait of the Solar System" has influenced many writers. One of the most poetic and beautiful treatments is C. Sagan, *Pale Blue Dot: A Vision of the Human Future in Space* (Random House, NY, 1994).

You can check out some of the early Gaia work in L. Margulis and J. E. Lovelock, "Biological modulation of Earth's atmosphere," *Icarus*, **21**, 471–489 (1974). David Attenborough, who directed programming for the BBC for many years, authored the very significant and beautiful *The Living Planet: A Portrait of the Earth* (Little Brown, Boston, MA, 1984).

A source for details regarding the stages of Gaia is E. Jantsch, *The Self-organizing Universe: Scientific and Human Implications of the Emerging Paradigm of Evolution* (Pergamon Press, NY, 1980).

To read about possible high-tech approaches to alleviate global climate change and related problems, check out our recent book: G. Matloff, C Bangs, and L. Johnson, *Harvesting Space for a Greener Earth*, 2nd edn. (Springer, NY, 2014).

Information on the Game of Life is available online (*http://en.wikipedia.org/wiki/Conway's_Game_of_Life*).

A very readable discourse on the conscious Gaia hypothesis is A. A. Wyller, *The Planetary Mind* (MacMurray & Beck, Aspen, CO, 1996).

Books dealing with the search for extrasolar planets become obsolete very quickly. A dated, albeit nice semi-technical introduction to this burgeoning field of research is S. Clark, *Extrasolar Planets: The Search for New Worlds* (Wiley–Praxis, Chichester, U.K., 1998). Another book of the same vintage is K. Croswell, *Planet Quest* (Free Press, NY, 1997). A more recent and more technical treatment is J. W. Mason, ed., *Exoplanets: Detection, Formation, Properties, Habitability* (Springer–Praxis, Chichester, U.K., 2008). A frequently updated online source is *The Extrasolar Planet Encyclopedia* (*http://exoplanet.eu/*).

The classic study of extrasolar planet habitability is S. H. Dole and I. Asimov, *Planets for Man* (Random House, NY, 1964). A very comprehensive review of the search for extraterrestrial life is S. J. Dick, *The Biological Universe* (Cambridge University Press, 1996).

One of the nearest known extrasolar planets is within the habitable zone of a red dwarf star. Its discovery was described in June 2014 in the online article at *file:///Users/gregmatloff/Desktop/books/Star%20Light,%20Star%20Bright%3F/Chapter%208/Nearby%20Alien%20Planet%20May%20Be%20Capable%20of%20Supporting%20Life.webarchive*

PART III

Fictional bright stars

The author and the poet look at stellar and universal consciousness

CHAPTER 9

The poet and the Cosmos

Ah Ben!
Say how, or when
Shall we thy guests
Meet at those lyric feasts
Made at the Sun,
The Dog, the Triple Tun?
Where we such clusters had
As made us nobly wild, not mad;
And yet each verse of thine
Outdid the meat, outdid the frolic wine.

Robert Herrick, *An Ode to Ben Jonson*

In his brilliant essay "The last magician," Loren Eisley quotes an essay of Emerson comparing the poet with a lightning rod. It is a nice analogy. Like the lightning rod, a poet must project toward the heavens while remaining rooted in the damp soil of the natural world. As has been speculated in an online essay about Sappho, one of the earliest lyric poets, poetry might have originally been a form of magic. By manipulating the words in a poem, the poet might gain some control over the object or subject described in that poem. Unfortunately, not much of Sappho's work has survived and it is difficult to ascertain her thoughts on stellar and universal consciousness.

However, at least some of the work of Athenian playwrights of the 5th century BC has survived. Among these is Sophocles (497–406 BC). In his Oedipus the King, the Chorus laments the fate of any individual who is not blessed by the Sun's (Apollo's) illumination:

Oh no! By God the Sun, who stands foremost in the heavens,
unblessed and accursed may I be, cast in utter darkness,
my lord, if I have any such thought!

THE ANCIENTS

At about the same time that Sappho lived on the Aegean island of Lesbos, scholars in the Levant were composing the Old Testament of the Bible. If you check through this document for evidence of stellar or universal consciousness, you will not find a great deal. Early Hebrew authors were more interested in the relationship of God to man than in whether attributes of organic life were shared by celestial objects. Like most ancients, they placed a stationary Earth at the center of the Cosmos. Moreover, perhaps to distance themselves from the astrology of their Babylonian conquerors, they pointedly rejected sky and nature cults.

However, in scriptural writings from the later Hellenistic period, there are exceptions to this generalization. One can be found in Verse 43 of Ecclesiasticus, which was originally written in Hebrew and translated into Greek in around 130 BC. This poem describes the glories of the heavens. Relevant portions are reproduced below:

The sun

Pride of the heights, shining vault,
  So, in a glorious spectacle, the sky appears
The sun, as he emerges, proclaims at his rising,
  "A thing of wonder is the work of the Most High!"
At his zenith he parches the land,
  Who can withstand his blaze?
A man must blow a furnace to produce any heat,
  the sun burns the mountains three times as much;
Breathing out blasts of fire,
  flashing his rays he dazzles the eyes.
Great is the Lord who made him,
  And whose words speeds him on his course.

The moon

And then the moon, always punctual,
  to mark the months and make division of time:
the moon it is that signals the feast,
  a luminary that wanes after her full.
The month derives its name from hers
  she waxes wonderfully in her phases,
banner of the hosts on high,
  shining in the vault of heaven.

The stars

The glory of the stars makes the beauty of the sky,
  a brilliant decoration to the heights of the Lord.
At the words of the Holy One they stand as he decrees,
  and never grow slack at their watch.

It is not surprising that this fragment implies that the heavenly hosts and the Lord inhabit the celestial realm. However, it is interesting that the apparent motion of the Sun requires a push from the deity rather than being a function of the nature of the Sun as most ancient philosophers suspected. The fact the stars could follow a holy decree implies a form of consciousness. The fact that the Moon is a "her" and the Sun is a "he" also implies a connection with earlier considerations of these objects as celestial deities. This passage must have been a challenge for literalists who later adopted it for the Christian canon.

Perhaps because of this short passage, ancient concepts of a conscious Cosmos did not disappear with the fall of the Roman Empire. They reappear many times as the writing of poetry and prose mature. We present a sampling of these writings in the following sections.

POETS OF THE MEDIEVAL WEST

During the Middle Ages, roughly between 500 AD and 1500 AD, poetry was one of the creative forms that flourished in Western Europe. Because of the influence of Christianity, most poems of this period dealt with the relationship between God and man, rather than with details of the celestial realm.

One significant poetic voice of this period is Dante Alighieri (1265–1321). Born in Florence (Italy) and deceased in Ravenna (Italy), Dante's most famous work is *The Divine Comedy*. In this monumental work, the narrator visits hell and heaven.

Like most medieval Western scholars, Dante placed the Earth in the center of the Universe with Jerusalem at the center of the Earth. Hell was near the center of the spherical Earth with purgatory above it. Above Earth's surface were the spheres of the geocentric solar system model. The Moon was closest to Earth, followed by Mercury, Venus, the Sun, Mars, Jupiter, Saturn and the "fixed" stars. Above this was a crystal sphere with Paradise above (Figure 9.1). Although the stars and planets were probably not deemed to be conscious beings, various attributes were assigned to the spheres of these celestial bodies. The Sun's sphere, for example, denoted wisdom.

The human role in this Universe is essentially that of a pilgrim. If we are virtuous enough, we may emerge through the uppermost sphere to join the heavenly host. Possibly, the view of life as a pilgrimage may be connected to the medieval phenomenon of the Crusades.

Another somewhat later poet with a medieval worldview is John Milton. Although technically living in Renaissance England (1608–1674), Milton preferred the geocentric model but mused in *Paradise Lost* about the possibility of the Sun being the center of the Solar System, rather than the Earth. This clearly places the poet at the intersection of the ancient geocentric and modern heliocentric worldviews. One quote from Milton,

> "The mind is universe and can make a heaven of hell, a hell of heaven,"

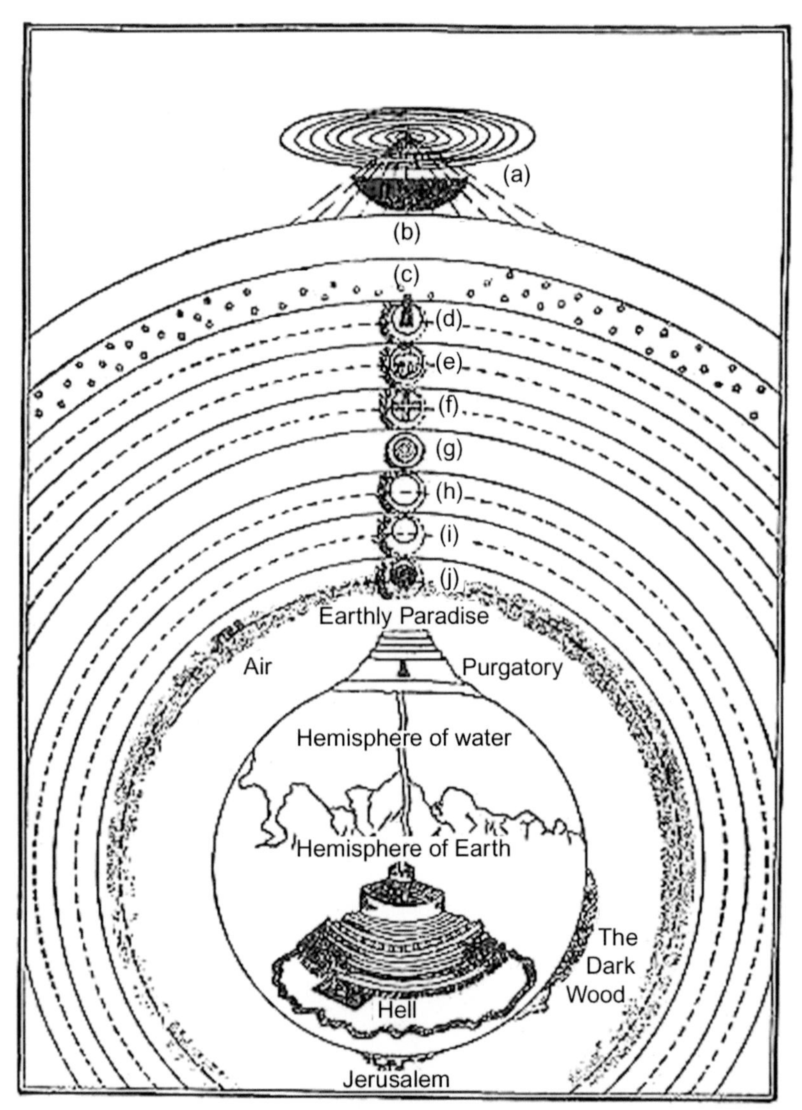

Figure 9.1. Dante's cosmology (from *The Divine Comedy*). (a) The Empyrean Paradise. (b) Crystalline Sphere, Primum Mobile. (c) Sphere of the Fixed Stars. (d) Sphere of Saturn. (e) Sphere of Jupiter. (f) Sphere of Mars. (g) Sphere of the Sun. (h) Sphere of Venus. (i) Sphere of Mercury. (j) Sphere of the Moon.

demonstrates that he did not think of the Universe as a mindless machine. In a schematic representation of Milton's Universe plotted on a two-dimensional sheet of paper, hell is at the bottom and heaven at the top. The geocentric solar system, with the Earth at center is in the middle of the page. Perhaps because of the unsettled politics of his day, Milton was greatly concerned that the Universe should act to preserve order and reduce the role of chaos.

THE BIRTH OF MODERNITY

The modern world can be thought of as arriving with the Renaissance. Clearly, the foremost poet (and playwright) of this period is William Shakespeare (1564–1616). With his vast output of plays and sonnets, Shakespeare did more than any other individual to explore and define the artistic limits of the English language. Like Milton, Shakespeare lived in a transitional age from the point of view of cosmogonies. In his youth, the geocentric solar system model of Ptolemy was still firmly in place. However, before his death, the telescope had been invented. In 1609, Galileo had pointed such an instrument toward the heavens. Anyone who could afford to purchase or grind two appropriate lenses and fit them in a tube could see for himself or herself that the Earth-centered conception of the Universe was becoming unglued.

Shakespeare appears to scoff at the practice of astrology and the doctrine that our lives are controlled by the positions of heavenly objects. I suspect that every graduate of secondary school in the English-speaking world, at least, had encountered these words from *Julius Caesar*:

> "The fault, dear Brutus, is not in our stars
> But in ourselves, that we are the underlings."

In *King Lear*, Edmund expresses a similar point of view. Astrology is here made to seem like an indulgence of ineffectual fools:

> "This is the excellent foppery of the world, that when we are sick in fortune,—
> often the surfeit of our own behavior,—we make guilty of our disasters
> the sun, the moon, and the stars: as if we were villains by necessity; fools by
> heavenly compulsion; knaves, thieves, and treachers, by spherical predominance;
> drunkards, liars, and adulterers by an unforced obedience of planetary
> influence; and all that we are evil in, by a divine thrusting on."

On the other hand, Shakespeare does seem to view humans as having a noble and cosmic destiny. This realization exists side by side with his surprisingly modern view of the ultimate fate of the Earth as the Sun expands in the far future to engulf it and all upon it, as evidenced in *The Taming of the Shrew*:

> "The cloud-capp'd towers, the gorgeous palaces,
> The solemn temples, the great globe itself,
> Yea, all which it inherit, shall dissolve,
> And, like this insubstantial pageant faded,
> Leave not a rack behind."

Although Shakespeare was certainly no fan of astrology, he was open to the possibility of the Sun's divinity. As Venus says to the Sun in the poem *Venus and Adonis*:

> "O thou clear god, and patron of all light,
> From whom each lamp and shining star doth borrow
> The beauteous influence that makes him bright . . ."

This passage demonstrates that astronomy at the time of Shakespeare had yet to prove that stars are distant Suns.

Even though the positions of the Sun, Moon, and naked-eye planets among the constellations could not rule our lives, Shakespeare did not rule out a more subtle celestial influence. This is demonstrated in his reference to tides in *The Winter's Tale*:

> "Swear his thought over
> By each particular star in Heaven and
> By all their influences, you may as well
> Forbid the sea for to obey the moon
> As or by oath remove or counsel shake
> The fabric of his folly . . ."

In William Shakespeare's vast body of work, there are many other celestial references. Some of these are discussed in the paper by W. H. Guthrie, which is cited below.

IN THE NEW LAND OF AMERICA

Walt Whitman (1819–1892) is foremost of the influential American poets with a cosmic consciousness (Figure 9.2). He worked as a poet, essayist, and journalist. Whitman is viewed as a humanist and is a transitional figure between transcendentalism and realism. Born in Long Island (New York), Whitman served as a volunteer nurse during the American Civil War. Between 1846 and 1848, he edited the *Brooklyn Eagle*. His influence on the then city of Brooklyn was enormous, as indicated by his poetry and his contribution to the creation of Fort Greene Park.

Much of his poetry is included in his monumental *Leaves of Grass*, which he began writing in 1850 and modified and edited until his death. Many of the poems in his book deal with the subject of astronomy. In his famous poem *When I Heard the Learn'd Astronomer*, Whitman displays an impatience with the scientific requirement for "proofs," "figures," "charts," and "diagrams". He leaves the lecture to sit in silence in the "mystical moist night-air" and looks up in "perfect silence at the stars."

Nevertheless, in his other astronomy-related poems, Whitman indicates an awareness that he has come full circle. Instead of admiring the Cosmos from

Figure 9.2. Walt Whitman (courtesy U.S. Library of Congress).

the viewpoint of a modern man, he participates in it like an ancient shaman. In *Song of Myself*, he writes:

"Swiftly arose and spread around me the peace and knowledge that pass all
the argument of the earth,
And I know that the hand of God is the promise of my own,
And I know that the spirit of God is the brother of my own,
And that all the men ever born are also my brothers, and the women my
sisters and lovers, ..."

"Immense have been the preparations for me,
Faithful and friendly the arms that have help'd me.

Cycles ferried my cradle, rowing and rowing like cheerful boatmen,
For room to me stars kept aside in their own rings,
They sent influences to look after what was to hold me."

Another very significant poem by Whitman is *Salut Au Monde!* Here is a fragment:

"Each of us inevitable,
Each of us limitless—each of us with his or her right upon the Earth,
Each of us allow'd the eternal purports of the earth,
Each of us here as divinely as any is here."

The above poetry fragments present a bit of Whitman's conception that his soul encompasses all creation. In *Song of the Open Road*, he implies that he shares consciousness with the stars:

> "The earth, that is sufficient,
> I do not want the constellations any nearer,
> I know they are very well where they are,
> I know they suffice for those who belong to them."

Decades before relativity, Whitman realized the significance of time and space. In *Gods*, he deifies them, as well as the Sun and stars:

> "Or Time and Space!
> Or shape of Earth, divine and wondrous!
> Or shape in I myself—or some fair shape, I, viewing, worship,
> Or lustrous orb of Sun, or star by night:
> Be ye my Gods."

Whitman's thought greatly affected many of his contemporaries. One such was the Canadian psychiatrist Richard Maurice Bucke (1837–1902). Burke's famous *Cosmic Consciousness*, a study of the mystical experience, expresses the opinion that human consciousness will evolve through a number of states, including loss of fear of death and sense of sin to a merger with the consciousness of the Cosmos. He became a friend of Whitman and credited Whitman with elevating him to a higher plane of being.

CONCLUSIONS: SETTING THE STAGE FOR STAPLEDON AND SCHRÖDINGER

There are many other poets and playwrights, of course, who trod the path toward an awareness of stellar consciousness. Many books the size of this one would be required to give them justice. However, what we see from the above is that a poet, Walt Whitman, speculates on space/time a half century before Einstein, and lives as if he shares consciousness with the entire Cosmos long before Schrödinger. It is possible that Bucke's *Cosmic Consciousness* was known to Olaf Stapledon, whose work is featured in the Chapter 10. As we will discuss, Stapledon's vision triggered the creativity of a host of scientists. His epochal masterpiece *Starmaker* may well be the first time that stellar and universal consciousness were published in a science fiction format.

FURTHER READING

Emerson is quoted in L. Eisley, "The last magician," *The Invisible Pyramid* (Scribner, NY, 1970). My source on Sappho is "Sappho, 620 BCE–550 BCE," from the Poetry Foundation (*http://www.poetryfoundation.org/bio/sappho*).

The quote from Sophocles is from an English version of *Oedipus The King*. It is published in J. Gassner, ed., *A Treasury of the Theatre: From Aeschylus to Turgenev* (Simon & Schuster, NY, 1951).

My biblical source is A. Jones, ed., *The Reader's Edition of the Jerusalem Bible* (Doubleday, Garden City, NY, 1968).

There are many online sources describing Dante's poetry and view of the Universe. One is "Dante's Cosmology: The Structure of Creation in Dante's Masterwork" (*https://medium.com/@litmuse/dantes-cosmology-7d61b69b3b59*). Figure 9.1 is also from a web source (*http://www.sacred-texts.com/earth/boe/boe27.htm*).

John Milton's poems and cosmology are discussed in many sources. A good introduction to *Paradise Lost*, attributed to Dartmouth College, is *http://www.dartmouth.edu/~milton/reading_room/pl/intro/index.shtml* The quote from Milton in the text is from *http://www.goodreads.com/quotes/715631-the-mind-is-a-universe-and-can-make-a-heaven* A representation of Milton's Universe, attributed to Walter Clyde Curry, can be found at *http://gypsyscholarship.blogspot.com/2010/01/miltons-cosmos-or-universe.html*

The article on Shakespeare and astronomy is W. H. Guthrie, "The astronomy of Shakespeare," *Irish Astronomy Journal*, **6**, 201–211 (1964); check out *http://articles.adsabs.harvard.edu/cgi-bin/nph-iarticle_query?1964IrAJ....6..201G&defaultprint=YES&filetype=.pdf*

Biographical data on Walt Whitman and Richard Maurice Burke were obtained from Wikipedia. The fragments of Whitman's poems used in this chapter and other chapters are from W. Whitman, *Leaves of Grass* (ed. Emory Holloway, Doubleday, Garden City, NY, 1926).

CHAPTER 10

A visionary named Olaf

When I dipt into the future far as human eye could see;
Saw the Vision of the world and all the wonder that would be.—

Saw the heavens fill with commerce, argosies of magic sails,
Pilots of the purple twilight dropping down with costly bales;

Alfred Lord Tennyson, *Locksley Hall*

Olaf Stapledon (1886–1950) certainly "dipt into the future" and glimpsed as far into the depths of time as is possible. His full name was William Olaf Stapledon. He earned B.A. and M.A. degrees from Balliol College, Oxford. After a brief stint as a teacher in Manchester, Stapledon worked as a shipping agent in Liverpool and Port Said from 1910 to 1913. A conscientious objector during the First World War, he served with the Friends Ambulance Unit. After receiving a Ph.D. in philosophy in 1925 from the University of Liverpool, he published his first book in 1929 on the subject of ethics.

Stapledon subsequently switched his medium to science fiction in the hope of reaching a wider audience. The success of his first book in this field, *Last and First Men*, published in 1931, encouraged him to adopt writing as a full-time career. The first edition of *Star Maker*, probably his most important novel from a scientific viewpoint, was published in 1937.

After World War 2, Stapledon lectured widely. He speculated on an interplanetary future for humanity at a British Interplanetary Society lecture in 1948 (Figure 10.1), promoted causes dealing with world peace and the anti-apartheid movement. His final book *Nebula Maker*, likely a first draft of *Star Maker*, was published posthumously in 1976.

Stapledon had a great many literary contacts. As Patrick Parrinder has reviewed, he corresponded with H. G. Wells, one of the founders of modern science fiction. As Liel Leibovitz writes online, Stapledon's Spinoza-like pantheism expressed in *Star Maker* encouraged C. S. Lewis to construct his famous Christian-based fantasy trilogy.

Although Stapledon was an agnostic, *Star Maker* is certainly significant from a spiritual point of view because of its conception of stellar and universal consciousness and the ultimate judgment of a fully conscious universe by the Star

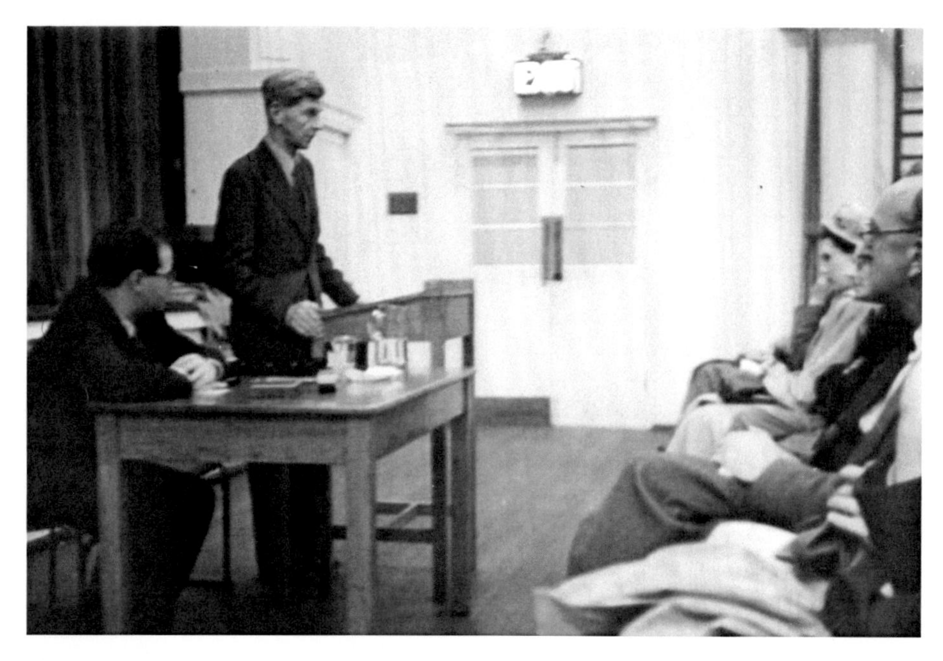

Figure 10.1. Stapledon addresses the British Interplanetary Society in 1948 (courtesy BIS).

Maker. It is also very significant because of the astronautic and scientific ideas that appear in its pre-war pages, perhaps for the first time.

STAR MAKER: THE NOVEL

Star Maker is a short novel, about 200 pages in length. Nevertheless, Stapledon masterfully portrays in it the evolution of the Universe from its creation (in a Big Bang–like event) to its ultimate interaction with the Creator of the Cosmos. Stapledon is able to accomplish all this in a novelistic setting by presenting a very human British narrator of the pre-WW2 era, who wanders on to the heath one evening after a small disagreement with his wife.

He meditates on the night sky and feels himself transported into space on a bodiless, astral journey. After traveling through space and time, he attempts to make contact with an alien species at the roughly contemporary human level of development. Until he contacts a philosopher who is amenable to sharing his mind, the intelligent aliens he encounters, fearing insanity, submit themselves to the "mental sanitation" authorities.

As a merged entity, the narrator and philosopher set off on additional interstellar journeys, merging with like-minded extraterrestrials at every destination. Because these mental mergers lead to an increase in consciousness, the narrator is able to experience extraterrestrial civilizations of increasing technological and psychological development as they progress through the galaxy.

One philosophical question investigated by Stapledon in *Star Maker* and other novels is "What is a human?" The narrator encounters near-humans, symbiotic organisms, hive creatures, living aquatic ships, intelligent plants, and other exotic forms. He comes to realize that it should be the response of an individual to the challenges of his/her environment and civilization that define "humanity", not the physical form.

As time goes on and their travels continue, the merged explorers witness examples of psychological evolution pointing in the direction of planetary minds, perhaps the precursors of Gaia. Some of these develop interplanetary travel and advance to the point that they begin exploration of their local galactic vicinities. In his talk presented at the BIS Stapledon Symposium, I. A. Crawford expresses the opinion that the establishment of an interstellar "community of worlds" might be the ultimate goal of space colonization.

In Stapledon's consideration of the future evolution of conscious species, he describes many sources of conflict, some of which doom promising civilizations. One is the disagreement between individualists and communalists, which played out in our time as the Cold War.

At this interstellar expansion phase of the novel, Stapledon considers that some of the communal minds directing ships of interstellar exploration become xenophobic in much the same way as a fraction of individual minds become insane. The advanced aliens they encounter must either be destroyed or absorbed. This sounds a lot like a 1937 version of *Star Trek*'s Borg culture.

Eventually, pacifist-minded worlds prevail and begin to build a galactic utopia. Two of the things the galactic peoples experiment with as they advance are disassembling giant planets and asteroids in planetary systems to create habitable shells around their stars and, therefore, accommodate a vastly increased population and diverting stellar trajectories to project entire solar systems on voyages of intergalactic exploration.

However, the stars obviously do not like this! Their reactions with planet-sterilizing filaments and nova-like eruptions demonstrate that stars do not like to be closed in and do not wish to have their paths around the galaxy's center altered. The minded planets establish contact with the stars and incorporate stellar consciousness into the growing galactic mind.

As it reaches out physically (and psychologically) to other galaxies, the galactic mind learns that spiral and elliptical galaxies themselves are conscious. Stapledon explored conscious galaxies more fully in his later, unfinished *Nebula Maker*. These entities and other galactic minds are incorporated into a growing universal intelligence.

All this time (which amounts to many billions of years), the developing cosmic mind has been experiencing hints of the Star Maker, the Creator of the Universe. In the novel's later stages, the cosmic mind finally encounters this infinite mind. What is learned by the cosmic mind is sad and inspiring. The purpose of creation is definitely not to support a platform for the development of individual souls who may be judged and redeemed by a benevolent creator. Moreover, the purpose of our Universe is not to develop a communal cosmic mind, composed of individual souls, who may join with the Creator in an ecstatic fashion.

Instead, our Cosmos is one of a series of developmental cosmoses. Because the cosmic mind evolved from our Universe is imperfect, the Star Maker examines it, notes its flaws, and figuratively puts our Cosmos back on the shelf. In this respect, the Star Maker is perhaps like an impersonal artist or curator rather than the personal god of many current terrestrial religions.

After its ultimate experience, the cosmic mind returns to our Universe, which is gradually dying. After this vast communal mind gradually disintegrates, the narrator finds himself standing once again in the British meadow, with no time elapsed.

The scientific predictions of *Star Maker*

Although *Star Maker* succeeds as a vision of a spiritual future, this does not explain its significance in scientific circles. Its significance to scientists lies in the quality of its 1937-vintage projections regarding science and technology. Some of these are discussed in the following subsections.

Direct mind-to-mind communication

Stapledon bases his thesis on the concept that telepathy will allow direct mind-to-mind communication and the eventual creation of planetary, interplanetary, galactic, and ultimately universal-linked minds. Although most scientists remain skeptical about achieving this goal through telepathy or other paranormal means, the World Wide Web is certainly a step in that direction.

In 2014 any human with access to a smartphone can readily tap into the knowledge base of humanity and communicate with billions of individuals anywhere on the planet plus a few not on the planet. As research in virtual reality produces nanoimplants, direct communication of one mind to another through this information network will almost certainly become possible at some point in the not very distant future. One wonders what the possibilities are for extraterrestrial civilizations millions of years in advance of us.

Genetic modification of life forms

In his consideration of advancing civilizations, Stapledon describes cultures that elect to modify their genetic structure to allow them to colonize other planets without the necessity for domed or underground cities. Humanity thus far has elected not to modify itself in such a manner; there are significant ethical, moral, and legal hurdles that must be overcome if such a path is followed in the future. However, biologists and agriculturists routinely alter the genetic structure of some of humanity's fellow travelers on the Earth. There is a great deal of controversy regarding these efforts. Stapledon would be very interested in the ongoing debate regarding genetic modification of terrestrial organisms.

Nuclear power

In the years preceding World War 2, many scientists were skeptical regarding its ultimate feasibility. Stapledon predicted in 1937 that advanced planetary civilizations would ultimately utilize mass–energy conversion to propel interstellar spacecraft. Since he lived to witness the nuclear destruction of Hiroshima and Nagasaki, he must have been amazed (and saddened) by the rapidity of "progress" in this field.

Cosmic impacts

Stapledon's narrator reports that some advanced civilizations are destroyed by cosmic impacts. When Stapledon composed *Star Maker*, the discovery that the demise of the dinosaurs about 65 million years ago was likely caused by the impact of a ~10 km diameter asteroid was decades in the future. The near-Earth asteroids, one of which recently impacted in Siberia and injured more that 1000 people, were also largely uncharted. Stapledon would be pleased with NASA plans to "lasso" one of these beasts and drag it in to a near-Moon orbit where astronauts can study it.

In-space construction: Worldships and space habitats

In his consideration of interstellar travel, Stapledon was one of the first to suggest the concept of the worldship: a spacecraft large enough to support a reduced-size planet-like ecology. We are a very long way from constructing worldships many kilometers long capable of supporting a population of a few hundred or thousand during millennial journeys to the nearest stars.

However, during the 1970s, a great deal of research was conducted in the U.S. and elsewhere on similar-sized stationary in-space habitats. These could be constructed in the near future, using near-term technology. With the addition of appropriate nuclear or solar engines, these craft could ultimately perform interstellar voyages of exploration and colonization. *Star Maker* is cited by many space habitat researchers.

In-space construction: Stapledon/Dyson spheres

Stapledon presents the case for a technologically advanced civilization disassembling planets and asteroids in its planetary system to construct a solid sphere around the parent star or a halo of many star-orbiting space habitats. This would allow a sizable population to efficiently tap the light emitted by the parent star and the resources of their planetary system. Freeman Dyson of the

Princeton Institute for Advanced Study has developed the concept mathematically and credits Stapledon with the original idea.

Researchers in the field of SETI (the Search for Extraterrestrial Intelligence) have begun to search nearby stars for signs of large-scale astro-engineering projects such as Stapledon/Dyson spheres.

Recently, a team at Oak Ridge National Laboratory in Tennessee coined the term "Dyson dot" for a huge solar reflector or occulter stationed so as to increase or decrease sunlight falling on Earth and, thereby, partially compensate for climate change.

The Big Bang theory of the Universe's origin

Long before the introduction of the Big Bang theory to scientific circles. Stapledon describes something very like it in *Star Maker*. He posits an initial eruption of matter, energy, space, and time. Moreover, he describes how our Universe cooled and expanded.

Modern cosmologists debate whether the Universe was created to optimize the chances of intelligent life evolving (the cosmological anthropic principle) or whether there exists a multiverse in which an enormous number of universes are constantly being created with the one we inhabit just happening, by random, to be acceptable for organic evolution. Stapledon's pantheistic cosmology incorporates aspects of both conceptions.

CONCLUSIONS: STAPLEDON'S APPROACH TO CONSCIOUSNESS AND THEOLOGY

In *Star Maker*, Stapledon posits an essentially pantheistic cosmology. Proto-consciousness permeates the Universe and can develop to an advanced state in nebula, stars, and organic life forms. The Creator of our Cosmos has little or no interest in the individual. There is no redemption of the soul, and there is no hell or heaven. To the Star Maker, it is only the finished product that counts. What is more, the universal mind, the ultimate achievement of our Cosmos, is merely a step in that direction. It is not surprising that C. S. Lewis created his famous fantasy trilogy as a Christian counterweight to *Star Maker*.

If you are an adherent of an Eastern religion, an atheist, agnostic, or pantheistic, you probably have no problem with Stapledon's theological speculations. However, if you are a monotheist, these concepts may be difficult to swallow.

Some may therefore be tempted to ignore Stapledon's philosophy and theology. Nevertheless, his vision has proven so accurate in scientific and technical circles. Because of the current dark matter crisis in astrophysics, in which about 70% of the matter in the Universe is missing but still affecting the motions of visible stars and galaxies, Stapledon's hypothesis of volitional stars cannot simply be ignored.

Chapter 11 discusses the influence of Stapledon's stellar conjecture on later generations of science fiction authors. Attention will then be turned to the astrophysics of the dark matter hypothesis (Chapter 12), stellar motions (Chapter 13), and a presentation of volitional stars as a scientific hypothesis (Chapter 14).

FURTHER READING

The biographical sketch on Olaf Stapledon presented above was derived from his Wikipedia entry (*http://en.wikipedia.org/wiki/Olaf_Stapledon*). A number of print biographies have also been published. For example, R. Crossley, *Olaf Stapledon: Speaking for the Future* (Liverpool University Press, Liverpool, U.K., 1994). This work is cited in S. Baxter, "Where was everybody: Olaf Stapledon and the Fermi paradox," *Journal of the British Interplanetary Society*, **65**, 7–12 (2012). A version of Baxter's paper was presented during the November 23, 2011 Starmaker Symposium at the British Interplanetary Society's Headquarters in London.

Stapledon's 1948 lecture to the British Interplanetary Society is reprinted as "Interplanetary man?" *Journal of the British Interplanetary Society*, **65**, 30–39 (2012).

My copy of *Star Maker* is O. Stapledon, *Last and First Men & Star Maker: Two Science Fiction Novels by Olaf Stapledon* (Dover, NY, 1968).

The poetic frontispiece for this chapter was inspired by its use by A. Sawyer, "The future and Stapledon's visions," *Journal of the British Interplanetary Society*, **65**, 25–29 (2012). Another paper presented at the British Interplanetary Society Stapledon Symposium is I. A. Crawford, "Stapledon's interplanetary man: A commonwealth of worlds and the ultimate purpose of space colonization," *Journal of the British Interplanetary Society*, **65**, 13–19 (2012).

Patrick Parrinder's paper on the interaction of Stapledon and Wells was delivered during the British Interplanetary Society Stapledon Symposium: P. Parrinder, "The Earth is my footstool: Wells, Stapledon and the idea of the post-human," *Journal of the British Interplanetary Society*, **65**, 20–24 (2012). Liel Leibovitz's online essay on the effects of *Star Maker* on C. S. Lewis is "Star Men" (November 1, 2011; *http://www.tabletmag.com/jewish-arts-and-culture/books/81969/star-men*).

My edition of Stapledon's final and posthomous work is *Nebula Maker & Four Encounters* (Dodd, Mead & Co., NY, 1983).

One source of information on how to prevent impacts from near-Earth asteroids is our recent book: G. Matloff, C Bangs, and L. Johnson, *Harvesting Space for a Greener Earth*, 2nd edn. (Springer, NY, 2014). This book also reviews concepts for in-space habitats. Two space habitat references that cite *Star Maker* are G. K. O'Neill, *The High Frontier: Human Colonies in Space* (Morrow, NY, 1976) and R. D. Johnson and C. Holbrow, *Space Settlements:*

A Design Study, NASA SP-413 (Scientific and Technical Information Office, NASA, Washington, D.C., 1977).

Work on nuclear-propelled and solar-propelled worldships is reviewed in G. L. Matloff, *Deep Space Probes*, 2nd edn. (Springer–Praxis, Chichester, U.K., 2005). Freeman Dyson discusses the contribution of Olaf Stapledon to his research on huge, artificial biospheres in F. Dyson, *Disturbing the Universe* (Harper & Row, NY, 1979).

The research on Dyson dots was conducted by a team consisting of Ken Roy, Robert Kennedy, David Fields, and Eric Hughes. Their work is presented and their technical papers cited in chapters 15 and 16 and appendices 3 and 4 of G. Matloff, C Bangs, and L. Johnson, *Harvesting Space for a Greener Earth*, 2nd edn. (Springer, NY, 2014).

The results of some of the thus-far unsuccessful searches for Dyson spheres have been published. See, for example, R. A. Carrigan, Jr., "IRAS-based whole-sky upper limit on Dyson spheres," *Astrophysical Journal*, **698**, 2075–2086 (2009). This paper is available online (*http://home.fnal.gov/~carrigan/infrared_astronomy/Fermilab_search.htm*).

CHAPTER 11

Stellar dreams in sci fi

> Thou orb aloft full-dazzling! thou hot October noon!
> Flooding with sheeny light the gray beach sand,
> The sibilant near sea with vistas far and foam,
> And tawny streaks and shades and spreading blue;
> O sun of noon refulgent! my special word to thee.
>
> Walt Whitman, *From Noon to Starry Night*, in *Leaves of Grass*

In the years following World War 2, a new generation of science fiction writers began to examine humanity's place in the Universe and other possible seats of consciousness. At least a few of them and their successors composed special words for the Sun and stars.

One of these is Arthur C. Clarke (1917–2008). Perhaps most famous for his post-war engineering studies of geosynchronous communication satellites and the novel that was the basis for the movie *2001, A Space Odyssey*, Clarke was a protégé of Stapledon and attended Stapledon's 1948 lecture at the London headquarters of the British Interplanetary Society (BIS).

STELLAR MUSINGS OF ARTHUR C. CLARKE

One of my first encounters with science fiction as an adolescent was a collection of short stories by Clarke. One of these, *Expedition to Earth*, was originally published in 1953. This is a classic example of a first-contact story and would certainly be considered science fiction rather than fantasy.

Expedition to Earth

Bertrond, one of the protagonists of this story, is a human-like extraterrestrial aboard a ship exploring our region of the galaxy. The time is approximately 100,000 BC. After decelerating from its interstellar velocity, the ship puts down near the Tigris–Euphrates rivers in what is today Iran or Iraq.

Under ordinary conditions, the crew of this exploration vessel and those that would follow, would educate the early humans they encounter to galactic standards within a few generations. However, these are not ordinary times. As Bertrond explains to Yaan, one of the humans they encounter, the Galactic Empire is dying. The ship will be recalled and further expeditions to our planet will not occur.

In their final meeting before the expedition departs, Bertrond communicates with Yaan and tries to warn of the dangers afflicting the Empire. Bertrond does this so that terrestrial humans in the far future may avoid the catastrophic mistakes leading to the Empire's destruction. The cause of this doom is the explosion of the Empire's stars.

As an adolescent, I could not understand what might cause the demise of the myriad stars composing an interstellar nation. However, after reading Stapledon's *Star Maker* in the 1970s, I realized that the Empire must have created Stapledon/Dyson spheres or manipulated stellar trajectories. And, as we know from *Star Maker*, either of these activities will make a star very, very angry!

Out of the Sun

Although I do not have a copy of this story in my science fiction collection, I have read it. Clarke must have been influenced by one of Stapledon's speculations expressed during his 1948 BIS lecture. Stapledon considered the possibility that one type of stellar organism might be a creature of fire and plasma optimized to thrive in the high magnetic fields and tremendous temperatures of the solar and stellar photosphere.

In *Out of the Sun*, Clarke recounts the experiences of a future human astronomer observing the Sun from an observatory on Mercury. He is amazed to obtain evidence of exactly the type of solar life considered by Stapledon.

Against the Fall of Night and *The City and the Stars*

In 1953, Clarke published a short novel, *Against the Fall of Night*. This was later expanded into *The City and the Stars*. The venue is Earth, about a billion years in the future. Humanity has fallen back to inhabit a single magnificent city and a rural district, connected by a tenuous underground rail line.

The young, city-born protagonist works to reunite the two population centers and also uncovers a lot of humanity's galactic history. He learns that earlier humans may have lost heart because of a psychic-spiritual problem brought about by the research activities of terrestrial and nonhuman scientists.

Two examples of disembodied sentience had been created: the Mad Mind and a more benevolent creature called Vanamonde. After a great deal of destruction to the surrounding Universe, the Mad Mind had been imprisoned in the heart of an artificial celestial object called the Black Sun. When that star dies in the far future, Vanamonde and the Mad Mind are destined to battle for universal dominance over the corpses of the dead stars. It is not impossible that this fictional imprisonment of a conscious being in the heart of an artificial star may have inspired Roger Penrose to speculate, as discussed in Chapter 6, that neutron stars may be conscious.

Other works by Arthur C. Clarke

Clarke has explored various aspects of consciousness in many other novels and stories. In his classic *Childhood's End*, he speculates that the ultimate fate of most sentient civilizations is to join with a nonmaterial consciousness spreading through the Cosmos. Those galactic civilizations that take the route of physically expanding through the galaxy are victims of stunted development.

In *2001, A Space Odyssey* and its sequels, Clarke expands upon Stapledon's concept of interstellar astral travel (which is also considered in *Childhood's End*) as a means of hastening the evolution of a human astronaut into a superhuman Star Child.

The famous astronomer, Carl Sagan, wrote one science fiction novel, *Contact*, which ultimately became a movie. The preparation of the human radio-astronomer heroine for her ultimate meeting with a citizen of the galaxy, applies a variation of Stapledon's/Clarke's astral travel. Not having a reliable theory for such a travel mechanism, Sagan falls back on the theoretically not impossible concept of a transgalactic hyperspace subway.

One of the sequels to Clarke's novel about an alien worldship that enters our Solar System, *Rendezvous with Rama*, features an interaction between the human heroine and the Star Maker, who is rather far removed from her problems in the story.

GREGORY BENFORD AND GORDON EKLUND: *IF THE STARS ARE GODS*

Greg Benford and his twin brother Jim are physicists who have contributed to NASA's ongoing efforts to launch solar sail probes toward the stars. Greg is also a successful and well-known science fiction author.

In *If the Stars Are Gods*, Benford collaborates with Gordon Eklund to expand an earlier short story of Eklund's. The thesis is very interesting: not only are stars conscious, but they possess god-like abilities as postulated by the ancients.

Advanced extraterrestrials arrive on Earth. However, they have only a passing interest in terrestrials. Stopping at the Moon, they elect to talk about their visit with a rather disgruntled and rather elderly former astronaut. They explain to him that their desire is to communicate directly with our Sun, which is apparently a benevolent being.

As the story advances, the reader learns a bit about the powers of the stars. One capability is to collect and permanently preserve selected organic life forms in a museum setting.

In other novels and stories, Benford has speculated on various aspects of consciousness. For example, he collaborated with David Brin on *Heart of the Comet* (Bantam, Toronto, Canada, 1986). In this novel, the authors present their concept of what one might experience while having one's consciousness uploaded into a computer.

FRED HOYLE: *THE BLACK CLOUD*

Fred Hoyle (1915–2001), one of the most influential astrophysicists of the 20th century, spent much of his career at the Institute of Astronomy at Cambridge. Although he contributed to various areas of astrophysics including stellar nucleosynthesis, he is the best known proponent of the steady state theory of cosmology, an alternative to the Big Bang theory, which is now almost universally rejected. In his consideration of nucleosynthesis, he supported the anthropic principle by arguing that the triple-alpha process by which carbon is created from helium in stellar interiors requires a specific resonance energy and spin to function and these requirements could probably not have developed at random. Hoyle also considered it unlikely that life originated on Earth, and favored instead the concept of panspermia, in which life spreads through space from living worlds to lifeless planets on which it can take root.

In 1957, Hoyle published a science fiction novel, *The Black Cloud*, which deals with conscious nebula. In this novel, astronomers in the early 1960s observe a dark celestial gas cloud or nebula on a trajectory directed toward our Solar System. It decelerates and takes up residence, blocking much of the sunlight reaching Earth. As global catastrophe looms because of rapidly falling terrestrial temperatures, astronomers succeed in establishing contact with this sentient molecular cloud.

Apparently, the cloud-mind had been unaware of the existence of planetary life. It agrees to move off from the Solar System after understanding the harm it was inadvertently responsible for. The most significant aspect of this novel might be its attention to the difficulty that diverse conscious forms in the Universe might have in recognizing and communicating with each other.

RAY BRADBURY: "HERE THERE BE TYGERS"

Ray Bradbury (1920–2012), whose formal education culminated with secondary school, was an American author who contributed to such fictional genre categories as horror, mystery, science fiction, and fantasy. One of his early literary influences was Edgar Allen Poe. In 1951 he composed the short story "Here There Be Tygers," which describes the interaction between human explorers and a planetary consciousness decades before Lovelock and Margolis proposed the Gaia hypothesis.

A human mining expedition lands on a beautiful, self-manicured park-like planet. The crew finds that if they think of flying, a gentle wind supports them aloft. If they feel hungry, schools of fish swim into hot springs and are cooked.

But the attempt of the mining supervisor to drill into the planet fails and his robotic drill is swallowed by the planet. When he considers more effective methods of defiling this paradise (including a nuclear weapon), he is killed, probably by a tiger-like predator manifested by the planetary intelligence.

After some debate, the ship's captain and most surviving crewmembers decide to leave the planet and lie to terrestrial authorities about its habitability to prevent future exploitation efforts.

As the ship rises, the crew notices that the vista has changed. Instead of a rolling parkland with fresh rivers tasting like wine, they observe volcanoes, dinosaurs, and tigers. As the ship departs the planetary system, the captain is informed that one member of the crew has elected to stay behind. They conclude that the harsh landscape is an optical illusion and that the missing crewmember will be coddled by this feminine-like planetary intelligence, which has spent billions of years in solitude and more than welcomes respectful human company.

In his career he revisited the theme of human explorers and colonizers conflicting with the inhabitants of other planets. This is particularly evident in his famous fantasy novel, the *Martian Chronicles* (1951), which evolved into a 1980 television mini-series.

I hope that *Martian Chronicles* operates as a cautionary myth. Perhaps because of this work and movies such as *Avatar*, humans will be more respectful of the life forms and ecologies they encounter if we someday expand to the planetary systems of other stars.

STANISLAW LEM: *SOLARIS*

The authors and scientists discussed in this chapter are either British or American and, therefore, have a decidedly Western slant on the concept of stellar or planetary consciousness. The same cannot be said regarding the Polish physician and author Stanislaw Lem (1921–2006). His classic science fiction novel *Solaris*, first published in 1961 and filmed in 1971, deals with the attempts of a human expedition to establish contact with a sentient ocean on the planet Solaris.

Hovering a few hundred meters above the surface of the living ocean, the scientists aboard the space station become increasingly frustrated with their unsuccessful attempts to contact this planet-wide oceanic brain that apparently spends its time conducting a theoretical monolog regarding the nature of the Universe. Contact is finally established as the ocean-mind creates from each person's subconscious a neutrino "phantom."

Lem seems to be commenting in *Solaris* on human attempts to establish contact with other conscious entities in the Universe. Rather than searching for nonterrestrial examples of consciousness, we search (in his opinion) for idealized, technically advanced versions of Earth. Before we can establish meaningful contact with extraterrestrials, we need to understand ourselves a great deal better than we do at present.

CONCLUSIONS: SOME OTHER EXAMPLES IN SCI FI AND FANTASY

As a visit to your local bookseller will reveal, the number of science fiction and fantasy books published each year is enormous. It is unfortunately impossible to sample all relevant titles in one short chapter. Nevertheless, here are two other book series worthy of consideration in this review.

The July 2014 issue of *Odyssey*, the online magazine of BIS, contains a short piece by Richard Hayes on stellar consciousness. As well as reviewing my presentation on the subject at BIS headquarters in June 2013, Stapledon's *Star Maker*, and Clarke's "Out of the Sun," Hayes discusses some recent novels by Iain M. Banks. Banks' *Surface Tension* and *The Hydrogen Sonata* feature, among other novelties, an extraterrestrial species called "field liners" who inhabit the magnetic field lines within stellar photospheres.

In the 1990s, fantasy writer Philip Pullman composed his award-winning "dark materials" trilogy, *Northern Lights*, *The Subtle Knife*, and *The Amber Spyglass*. The first of these multiple-reality novels became a Hollywood movie. In the second book of the trilogy, an astronomer from Earth learns of a connection between dark matter and consciousness. The dark matter mystery and the lack of success in the many attempts to solve it are the subject of Chapter 12.

FURTHER READING

My rather dog-eared copy of *Expedition to Earth* is one of 11 short stories in A. C. Clarke, "Expedition to Earth" (Ballantine, NY, 1953). According to Wikipedia, "Out of the Sun", which was originally printed in 1957, is reprinted in *The Collected Stories of Arthur C. Clarke* (Gollancz, London, 2001).

After appearing in the November 1948 issue of the magazine *Startling Stories*, Clarke's first novel *Against the Fall of Night* was first published in book form by Gnome Press (NY) in 1953. The expanded version, *The City and the Stars*, was published in 1956 by Frederick Muller (London).

Childhood's End was first published in 1953 by Ballantine Books (NY). The first edition of *2001, A Space Odyssey* was published in 1968 by Hutchinson Books (U.K.). The first of the "Rama" series, *Rendezvous with Rama*, was first published by Gollancz, London, in 1972. Carl Sagan's *Contact* was first published in 1986 by Simon & Schuster (NY).

I have been fortunate to locate a copy of G. Benford and G. Eklund, *If the Stars Are Gods* (Putnam Berkley, NY, 1977) in a bookshop specializing in used books. For an online review of this book by Trent Walters check out *http://www.sfsite.com/00a/is406.htm*

The biographical information on Fred Hoyle was taken from Wikipedia (*http://en.wikipedia.org/wiki/Fred_Hoyle*). In 2010, Penguin Classics (NY), issued a reprinted edition of *The Black Cloud.* Some of Hoyle's concepts on the possibility of prebiological or biological evolution in interstellar dust clouds are discussed in scientific papers such as N. A. Wickramsinghe, F. Hoyle, S. Al-Mufti, and D. H. Wallis, "Infrared signatures of prebiology—or biology?" *Astronomical and Biochemical Origins and the Search for Life in the Universe*, pp. 61–76 (eds. C. B. Cosmovica, S. Bowyer, and D. Werthimer, *Proceedings of the 5th International Conference on Bioastronomy, July 1–5, 1996, Capri, Italy, IAU Colloquium No. 161*, Editrice Compositori, Bologna, Italy, 1997).

Ray Bradbury's "Here There Be Tygers" was first published in 1951, but it has appeared in a number of science fiction anthologies since then. See, for example, *http://www.motherearthnews.com/nature-and-environment/ray-bradbury-here-there-be-tygers-zmaz78jfzgoe.aspx?PageId=1#axzz37I9pJQIK* (courtesy of *Mother Earth News*). The biographical information on Bradbury was taken from Wikipedia (*http://en.wikipedia.org/wiki/Ray_Bradbury*).

You can learn more about Stanislaw Lem from his website (*http://www.lem.pi*). Much scholarly attention has been devoted to S. Lem, *Solaris* (Walker, NY, 1961). One excellent commentary is by U.S. astronomer and historian of science S. J. Dick, *The Biological Universe: The Twentieth Century Extraterrestrial Life Debate and the Limits of Science*, pp. 258–259 (Cambridge University Press, NY, 1996).

PART IV

The astronomer's search

The astronomy and astrophysics of stellar and universal consciousness. What is the evidence?

CHAPTER 12

A dark mystery

When I heard the learn'd astronomer;
When the proofs, the figures, were ranged in columns before me;
When I was shown the charts and diagrams, to add, divide, and measure them,
When I, sitting, heard the astronomer, where he lectured with much applause in the lecture-room,
How soon, unaccountable, I became tired and sick;
Till rising and gliding out, I wander'd off by myself,
In the mystical moist night-air, and from time to time,
Look'd up in perfect silence at the stars.

Walt Whitman, *When I Heard the Learn'd Astronomer,*
from *Leaves of Grass*

Walt Whitman would probably approve of the arguments presented in this book for stellar consciousness. We have examined the history of the concept from the viewpoint of the mystic, the shaman, the astrologer, the philosopher, the poet, and the novelist. However, he would likely be uncomfortable with the necessity of attempting to move this concept from the realm of mysticism and fiction to the realm of speculative science. But, such an attempt is necessary in the modern world. I will do my best to keep the equations, charts, and tables to a minimum. If you have experienced secondary school physics, you should be able to follow the arguments. But, just in case, I have included some review concepts and conversions in the Appendices.

This chapter considers the mystery of dark matter. Stars within galaxies and galaxies within clusters do not move according to the predictions of Newton's familiar theory of universal gravitation. Is it possible that the anomalous motions are caused by some form of exotic material that has thus far eluded detection? Does Newtonian gravity theory require revision despite the fact it works so well within our Solar System? Or is some more exotic concept, such as volitional stars, necessary to solve this problem. The mystery of dark matter is a major problem: assuming that Newton's gravitational theory is absolutely correct and that some form of exotic, invisible matter is necessary to explain galactic and stellar motions, at least 65% of the Universe's mass is composed of this material. The current crisis in astrophysics caused by this apparent mass deficit may be as significant as the problems in the late 19th century that led to the rise of quantum mechanics and relativity.

One thing that the reader should realize is that astronomy, like all the sciences, is essentially a conservative endeavor. The reason many astrophysicists have joined the search for dark matter is that such an unseen mass has been invoked successfully in the past to explain anomalistic motions. Since the approach has succeeded before, maybe it would work again.

UNSEEN MATTER IN NEWTON'S SOLAR SYSTEM

The year is 1781. British astronomer William Herschel was searching the skies and charting faint stars with his telescope. He came across an unusual greenish object that he initially thought was a comet or nebula. However, further observation revealed that the object presented a visible disk when viewed through the eyepiece of his small telescope. Moreover, it orbited the Sun too slowly to be a comet. Furthermore, its orbit was close to the ecliptic, the path that the planets follow as they orbit the Sun. Herschel had discovered Uranus, the seventh planet out from the Sun and the first to be discovered in more than 3000 years.

Other astronomers began to study the newly discovered giant planet through their telescopes. It was not possible to fit an elliptical solar orbit to the observed position of the planet. After 50 years or so, the discrepancy between observation and theory was substantial enough that observational errors could be ruled out. So, it was proposed that an eighth Solar System planet existed beyond Uranus and that the gravitational pull of this object was the cause of the observed discrepancy in the solar orbit of Uranus. In 1845 the British astronomer John Adams was able to predict the approximate mass and location of the new planet. A few months later, the French mathematician Urbain Leverrier corroborated these predictions. In September 1846, the German astronomer Johann Galle discovered the giant planet Neptune close to the predicted position.

The discovery of Pluto, which is now considered to be a dwarf planet, is almost a replay of this process. Late in the 19th century, observational data seemed to suggest that the solar orbits of both Uranus and Neptune were perturbed by another planet farther out from the Sun. The American astronomer Percival Lowell commenced a search for this object in the early 20th century. In 1930, Pluto was discovered by American astronomer Clyde Tombaugh, utilizing improved photographic techniques.

Although it now seems likely that the supposed irregularities in the solar orbits of the two outermost planets do not really exist, the mathematical solutions to the equations of celestial mechanics did lead to the search for Pluto. It was a lucky break that this small, dim, low-mass world was close enough to the predicted sky position of the invisible, perturbing object that it could be observed.

DARK MATERIAL GOES STELLAR AND GALACTIC

During the first decades of the 20th century, observing technology had improved to the point that astronomers could determine the motions of stars around the center of our galaxy and estimate the masses of other galaxies. They also learned that galaxies tend to reside in clusters maintained by the mutual gravitational attraction of galaxies.

One of the scientists studying galaxy clusters was the Swiss astronomer Fritz Zwicky (1898–1974), who was affiliated to the California Institute of Technology. In 1930, he coined the term "dark matter" to explain the fact that the gravitational mass of the Coma Cluster of galaxies was considerably greater than the mass of luminous material in that cluster. In recent years, space observatories have begun to find evidence of some form of nonluminous material within clusters of galaxies (Figure 12.1).

Vera Rubin is another American astronomer who has contributed to the study of dark matter. The first woman to use the equipment at Mount Palomar (California) in 1965, she concentrated her studies on the angular rotation rate of galaxies. Our consideration of the dark matter phenomenon concentrates on stellar motion around the centers of host galaxies rather than intergalactic dark matter.

Astronomers had expected that the revolution of stars around the centers of host galaxies would follow a similar pattern to the Keplerian motion of planets in our Solar System. In the Solar System, planets closer to the Sun tend to move faster than those farther out. Mercury is about 40% the distance of Earth from the Sun. It revolves around the Sun at an average velocity of about 48 kilometers per second. Earth's solar-orbital velocity is approximately 30 kilometers per second. The distance between Mars and the Sun is about 50% larger than the average Earth–Sun distance (called the astronomical unit, or AU). Mars orbits the Sun at about 24 kilometers per second. Instead, they noticed that stars farther from the centers of their galaxies tend to move faster than those closer in. It is almost as if the stars in a galaxy are attached to the spokes of an invisible wheel. It is believed spiral galaxies like our Milky Way are able to maintain their shapes over multibillion-year timescales.

SOME ASTROGEOGRAPHY

Before going further, it is a good idea to try to grasp something of the distances we are talking about. Space is vast. The Earth orbits the Sun at an average distance of about 93 million miles or 150 million kilometers (1 AU).

Every year, a light photon in the vacuum of space travels about 60,000 AU at a velocity of 300,000 kilometers per second (186,300 miles per second). This distance is called 1 light year. The nearest stellar neighbors to our Sun, the three stars of the Alpha/Proxima Centauri system, are at a distance of about 4.3 light years. Figure 12.2 presents a Hubble Space Telescope image of a spiral

Figure 12.1. Hubble Space Telescope image of galaxy cluster ZwC10024 + 1662. The dark ring may be dark matter resulting from two galaxy clusters colliding (courtesy NASA).

galaxy much like our Milky Way. The Milky Way is about 100,000 light years across.

Spiral arms contain billions of stars as well as the dust and gas from which young stars form. Old stars reside in the central hub. Some old stars are also found in globular clusters, which enclose the galaxy in a spherical halo. Each globular cluster contains about 100,000 stars. At the very center of most or all spiral galaxies is a black hole millions of times as massive as the Sun. Our Solar System is about halfway out from the center of the Milky Way, in the rarefied interstellar medium between spiral arms. There are approximately 300 billion stars in a typical spiral galaxy.

The Universe contains a few hundred billion galaxies. Many of these are spirals, others are elliptical (roughly egg shaped), and some are cloud-like irre-

Figure 12.2. Hubble Space Telescope image of spiral galaxy M101 (courtesy NASA).

gular galaxies. The Universe, which formed in the Big Bang about 13.7 billion years ago, is perhaps 20 billion light years across and contains approximately 100 billion galaxies.

NEWTON'S SHELL THEOREM

Over the past seven decades, astronomers have proposed a number of possible solutions to the dark matter paradox. There is, of course, considerable interest in attempting to locate an apparent deficit of perhaps 70% of the Universe's mass. One tool used in investigating stellar revolution rates about the galactic center is a theorem familiar to students who have taken university level physics. This is Isaac Newton's shell theorem.

To understand how this theorem works, first realize that the gravitational attractive force between two separated objects is proportional to the product of the masses of the objects and inversely proportional to the square of the distance between their centers. In the case of the Earth orbiting the Sun, the separation that counts is the distance between the Sun's center and the Earth's center.

Now consider the case of a planet orbiting its star at a distance of 10 million kilometers in a perfectly circular orbit. If the planet is in a very dusty solar system, the shell theorem states that, when you calculate the properties of the planet's orbit, you can ignore dust farther from the planet's central star than the planet's separation from that star. Dust closer than 10 million kilometers can be treated when considering the planet's orbital properties as if it is concentrated at the star's center. When calculating the orbit of a planet farther out, say at 20 million kilometers, the central star will appear more massive in the calculations because all the dust out to the second planet's more distant orbital position must be treated as if it is located at the star's center.

The calculations become more complicated when considering a real celestial object such as a planet, asteroid, or comet. This is because orbits for real objects are usually elliptical. In our dusty Solar System, a celestial object "sees" a more massive central star when it is farther from that star than when it is closer.

MACHOS: THE CASE FOR AND THE CASE AGAINST

Faced with the dark matter crisis, some astrophysicists have fallen back on the successful strategy that led to the discoveries of Neptune and Pluto. Since the motion anomaly is more significant in the outer portions of galaxies and deep in intergalactic space, perhaps some form of physical object in galactic haloes is the culprit. One approach has been to search for MACHOS (massive astrophysical compact halo objects). Candidate MACHOS include subluminous but massive white dwarf stars, neutron stars, and black holes.

Although MACHOS are a logical choice, there are major problems with invoking them to account for dark matter. One problem involves globular clusters, such as M13 in Hercules (Figure 12.3), which surround the Milky Way and other galaxies in a spherical halo. Each globular cluster contains hundreds of thousands or even millions of old stars.

If there were enough MACHOS drifting around in galactic haloes to account for dark matter, there would be some disruption to the spherical shapes of these clusters. This has never been observed.

Another problem is gravitational focusing. If a massive, celestial object occults a more distant luminous celestial object, the gravitationally warped space-time near the closer object acts somewhat like a lens (Figure 12.4). Astronomers then observe an amplified but distorted image of the more distant object. If MACHOS were a major contributor to dark matter, the celestial gravitational lens effect would be more common.

WIMPS are very strong: If they exist!

Not too encouraged by the prospect for MACHOS, many astrophysicists and particle physicists favor matter at the other end of the size spectrum, WIMPS

Figure 12.3. A Hubble Space Telescope image of globular cluster M13 (courtesy NASA).

(weakly interacting massive particles). In this approach, the visible Universe is immersed in a sea of invisible subatomic particles with no electric charge which only react with the visible Universe according to Newton's law of universal gravitation. To account for the dark matter anomaly, WIMPS must account for about 70% of the Universe's mass.

There are problems with WIMPS, however. First, assume that WIMPS are slow moving with respect to the speed of light, or "cold." Since the gravitational attraction between WIMPS and normal matter must be mutual, we would expect pools of WIMPS to gather around massive objects such as stars.

Searches for cold WIMPS have been conducted by applying the shell theorem discussed above. The trajectory of an object in a slightly elliptical orbit like a planet or asteroid, a nearly parabolic orbit like a comet, or a hyperbolic orbit like the Pioneer 10/11 or Voyager 1/1 extrasolar probes would change as

Figure 12.4. Hubble Space Telescope image of galaxy cluster Abell 1689 gravitationally lensed (courtesy NASA).

the solar separation changes because of the WIMP cloud. No such effect has been noticed. In a recent study by N. P. Pitjev and E. V. Pitjev published in *Astronomy Letters*, the maximum WIMP mass within the solar orbit of Saturn (at 9.5 AU) is less than 1% of the Moon's mass. As you may recall, the Moon is about 1/81 the mass of the Earth. Such a small cold WIMP density in the Solar System, which is born out in other studies, renders the cold WIMP hypothesis less tenable.

Another alternative is "hot" WIMPS: WIMPS that move around the Universe at such high velocities that they cannot be constrained by stars. This hypothesis too has problems. First, astronomers have known for decades that the motions of stars in the inner or mid ranges of spiral galaxies (such as the Sun) move according to the familiar laws of Isaac Newton and Johannes

Kepler. You can check this by looking at fig. 23.21, on p. 637 of the 6th edition of the Chaisson/McMillan textbook cited in this chapter's "Further reading." The same curve is also in the first edition of that text, which was published in 1993. A very similar curve is published as fig. 26.16 on p. 467 of James B. Kaler, *Astronomy!* (HarperCollins, NY, 1994). This apparent lack of hot dark matter in the galactic vicinity of the Sun is corroborated by a 2012 study by C. Moni Biden and colleagues.

This is not conclusive, of course. However, it does at least seem possible that hot WIMPS avoid the interior of galaxies, as well as being invisible, massive, uncharged, and nonreactive with ordinary matter. Another problem for WIMP theoreticians is determining the acceleration mechanism for these hypothetical massive and uncharged particles.

The search for hot WIMP candidates continues at such places as the Large Hadron Collider at CERN, in Switzerland. It is not impossible that future interstellar space probes, operating at velocities much faster than those currently feasible, might resolve the problem if trajectory deviations are observed (or not observed) as the spacecraft moves farther out into interstellar space.

OF MAGNETISM AND MONDS

There are at least two other pieces in the dark matter jigsaw puzzle. One is magnetism. Much of the luminous matter in the Universe is located deep in stellar interiors and is in a heavily ionized state. Electrons and ions are affected by magnetic fields. Moreover, a galactic magnetic field exists as well. Could this field be a contributor to the anomalous stellar motions attributed by many to dark matter. According to a 2004 study published in *Ap. J.*, probably not.

Another possibility is that our assumptions regarding Newton's and Einstein's gravity theories may not be exactly correct over galactic or cosmological distances. Vera Rubin, for one, acknowledges the possibility that it may be necessary to derive a modification of Newtonian dynamics (MONDS). Why, for example, must the gravitational attraction between two massive objects vary exactly with the inverse square of the distance between their centers? Is it possible that the inverse exponent may be 1.9999 or 2.0001 instead of exactly 2? But as noted in a 2011 paper by S. Capozziello and colleagues, different modifications to Newton's gravitational law seem to be necessary to explain kinematics anomalies at different distances.

CONCLUSIONS: MULTIPLE CAUSES?

It seems as if there are no easy solutions to the problem of anomalous stellar and galactic motions. WIMPS, MACHOS, magnetism, and MONDS have all been suggested and considered. But, what if there are multiple causes? The suggestion put forth in this book is that there may be a different cause behind stellar motions in the outer fringes of spiral galaxies and galactic cluster effects. Perhaps stellar volition can be applied to average star motion within galaxies and a modification of Newton's law of universal gravitation applied to explain galaxy clustering. Chapter 13 presents observational evidence that stars cool enough to have molecules in their photospheres move differently than hotter stars.

FURTHER READING

The story of the discovery of Uranus, Neptune, and Pluto is told in many places. A good reference is the college textbook by E. Chaisson and S. McMillan, *Astronomy Today*, 6th edn. (Pearson Addison-Wesley, San Francisco, CA, 2008).

Fred Zwicky's contributions to the concept of dark matter and other aspects of observational astronomy are available online (*http://en.wikipedia.org/wiki/Fritz_Zwicky*). Information on Vera Rubin's contributions is available online (*http://en.wikipedia.org/wiki/Vera_Rubin*).

Data on Solar System planet orbits are from tables in K. Lodders and B. Fegley, Jr., *The Planetary Scientist's Companion* (Oxford University Press, New York, 1998). Galactic and universal data are from the above-cited reference by Chaisson and McMillan. Data on MACHOS and WIMPS come from the same source.

Most university level physics texts consider Newton's shell theorem. A good source is H. C. Ohanian, *Physics*, 2nd edn. (Norton, NY, 1989).

A recent observational study for cold dark matter within the Solar System is N. P. Pitjev and E. V. Pitjev, "Constraints on dark matter in the Solar System," *Astronomy Letters*, **39**, 141–149 (2013). This study was based on 677,000 positional observations of planets and spacecraft.

The search for dark matter within the Sun's galactic vicinity mentioned in the text is C. Moni Biden, C. Carraro, R. A. Mendez, and R. Smith, "Kinematical and chemical vertical structure of the galactic disk, II: A lack of dark matter in the solar neighborhood," *Astrophysical Journal*, **747**(101), 13 (2012), *http://arxiv.org/1204.3924v1*

I have recently discussed the possibility of searching for hot WIMPS by applying Newton's shell theorem to examine the trajectories of very-high-velocity interstellar spacecraft: G. L. Matloff, "Extrasolar sail trajectories and dark matter," *Acta Astronautica*, **104**, 472–476 (2014), *http://dx.doi.org/10.1016/j.actaastro.2014.03.019*

One study demonstrating that magnetism is not a good candidate to account for the anomalous stellar motions attributed by many astrophysicists to dark matter is F. J. Sanchez-Salcedo and M. Reyes-Ruiz, "Constraining the magnetic effects on HI rotation curves and the need for dark halos," *Astrophysical Journal*, **607**, 247–257 (2004).

An online archived paper discussing various aspects of the dark matter puzzle including attempts to modify Newton's law of universal gravitation is S. Cappozziello, L. Consiglio, M. De Laurentis, G. De Rosa, and C. Di Donata, "The Missing Matter Problem: From the Dark Matter Search to Alternative Hypothesis," *arXiv:1110.5026v1* [astro-ph.CO], October 23, 2011.

CHAPTER 13

A kinematics anomaly

> The one who lets down Cassiopeia's hair
> And feeds Draco whenever he will eat
> And binds the sandals on the sturdy feet
> Of the Herdsman, and who chains the glittering Bear
>
> Merrill Moore, *Constellation*

As I discussed in the "Introduction," I did not enter this research process anticipating that Olaf Stapledon's concept of stellar volition would have any observational support. But I was wrong! At least in the Sun's galactic vicinity, those stars that are cool enough to have molecules move on average a bit faster than their hotter cousins in their orbits around the galaxy's center. Perhaps the sandals on the feet of the Herdsman are not bound uniformly!

Before we begin considering this discontinuity in star velocity, it is necessary to review how astrophysicists classify stars. Since there are billions of stars in the Milky Way, this is done in a manner not dissimilar to techniques used to classify human populations.

STELLAR CLASSIFICATION AND THE HERTZSPRUNG–RUSSELL (H-R) DIAGRAM

Imagine that you have the job of classifying all the people in a large city. You would do this by conducting a survey to gather information on height, weight, gender, ethnicity, age, and a host of other factors. Then, you would coordinate and arrange the data in a graphical form for easy interpretation.

In the early decades of the 20th century, astronomers were able to perform the same task for hundreds of thousands of stars in our galaxy. This is because the U.S. Civil War had resulted in great advances in the art of optics. A new generation of large refracting telescopes, many of which still exist, became available at U.S. university observatories in the decades after that war. Many were equipped with mechanical clock drives allowing newly developed cameras at the telescope eyepiece to photograph celestial objects. With the aid of prisms, stellar spectra could be routinely imaged at the same time.

By photographing star positions at 6-month intervals, distances to the nearest stars could be estimated with increased accuracy. Observation of binary stars in orbital planes allowing for periodic eclipses yielded information on star sizes relative to the Sun.

It was necessary to find a class of astronomical workers who could gather, reduce, and correlate all these data. In the U.S., this was accomplished by young women such as Annie Jump Cannon and Cecilia Payne-Gaposchkin, who analyzed the spectra of hundreds of thousands of stars while enduring the demeaning title of "computers" and a much lower salary than their male counterparts. Around 1920, these data were organized into the Hertzprung–Russell (H-R) diagram by the Danish astronomer Ejnar Hertzprung and the U.S. astronomer Henry Norris Russell (Figure 13.1).

To appreciate the basics of stellar classification, first look at the left vertical axis. This lists stellar luminosity (or radiant output) relative to the Sun. The dimmest stars emit 10^{-4} (1/10,000) as much electromagnetic radiation as the Sun; the brightest emit 10^4 (10,000) times as much electromagnetic radiation as the Sun.

The right vertical axis presents the same information using the absolute magnitude scale. The brightest stars in the evening sky have an apparent visual magnitude of about 0; the dimmest stars visible to the unaided human eye have an apparent visual magnitude of around 6. When a star's apparent visual

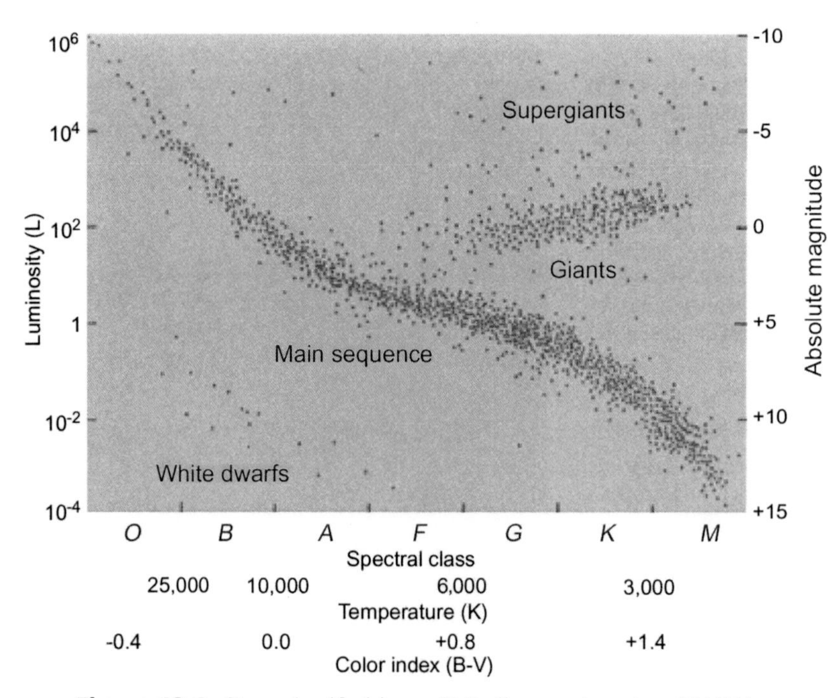

Figure 13.1. Stars classified in an H-R diagram (courtesy NASA).

magnitude decreases by 1, say from 5 to 4, the electromagnetic radiation we receive from it increases by about 2.5×. In the absolute magnitude scale, all stars are mathematically moved the same distance from the Solar System (about 32.6 light years). With distance removed in that manner, stellar radiant outputs can be compared.

Now look at the first line on the horizontal axis. Star spectral classes, starting from left to right, are O, B, A, F, G, K, and M. These spectral classes are subdivided further: for example, B stars are classified B0 through B9.

Although the interior of a hydrogen-fusing star such as the Sun has a temperature of about 20 million degrees Kelvin, the surface temperature of stars vary with spectral class. The hottest O-class stars radiate at around 30,000 degrees Kelvin; the coolest M-class stars radiate at around 3000 degrees Kelvin. Our Sun is a G2-class star with a surface temperature a bit less than 6000 degrees Kelvin.

Star color and mass also vary with spectral class. The most massive stars, with perhaps 50æ solar mass, are blue O stars. The least massive, with less the 10% the mass of the Sun, are red M stars.

Most stars, including the Sun, are on the main sequence. Hot, blue O stars reside on the main sequence for "only" a few million years. Our G2-class Sun is about halfway though its main sequence life expectancy of 10 billion years. Some M-class red dwarf stars will still be on the main sequence in a trillion years.

The bottom horizontal scale in Figure 13.1 is the B-V color index. This relates the luminosity of a star in the blue region of the visible light spectrum to its luminosity in the yellow–green region of the visible light spectrum.

STELLAR EVOLUTION

A star begins its life as a protostar: a growing clump of matter contracting within its birth nebula (Figure 13.2). As gas and dust strike the infant star, it begins to glow. A typical Sun-like protostar enters the H-R diagram when its surface temperature reaches about 3000 degrees Kelvin and its luminosity is about 100æ that of the present-day Sun.

Gradually, the star contracts and increases in surface temperature. When it reaches the lower boundary of the main sequence and a surface temperature of about 6000 degrees Kelvin, the density, pressure, and temperature near the star's core are sufficient for hydrogen fusion to begin. The star has now turned on as a main sequence star and is converting hydrogen into helium and energy.

During the next 10 billion years, the star gradually increases in luminosity as it climbs toward the upper boundary of the main sequence. As is true for all main sequence stars, our Sun's luminosity is slowly increasing. It is now about midway between the lower and upper boundaries of the main sequence. In about a billion years, unless terrestrial life takes some preventive measures, the increased luminosity of the Sun may cause the Earth's oceans to boil away.

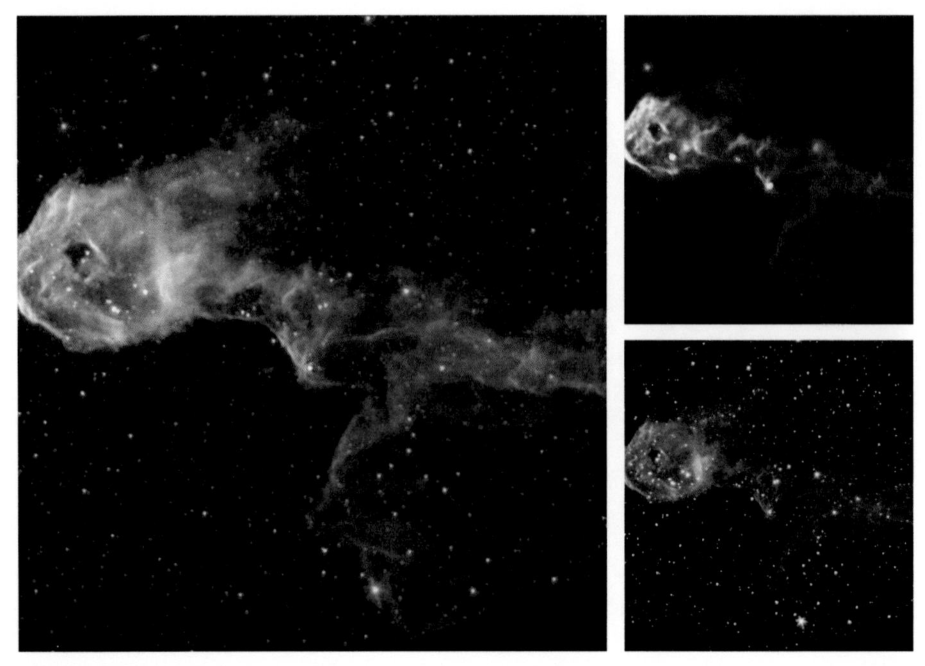

Figure 13.2. Multispectral images of protostars forming within stellar birth nebulae IC 1396 (courtesy NASA).

When an aging Sun-like star (or less massive star) leaves the main sequence, it expands and cools to become a giant star. Now, it burns helium to form beryllium and carbon as well as fusing hydrogen. Carbon acts as a catalyst to greatly speed up the hydrogen–fusion reaction rate. Within another hundred million years or so, the giant star has used up most of its fusion fuel and begins to contract. It crosses the main sequence and continues toward the lower left corner of the H-R diagram, finally becoming a white dwarf. These hot, subluminous objects compact the mass of a star (perhaps 300,000 times the Earth's mass) into an object the physical size of the Moon. Atoms in white dwarfs are so close together that there is no space between them. Electrons are shared by atomic nuclei in these stars.

High-mass stars (greater than 8 solar masses) evolve faster and eventually become supergiant stars. These are element factories in which a variety of fusion reactions create atoms as massive as iron. As fusion fuel runs out, the supergiant rapidly contracts and rebounds as a supernova. In this titanic explosion, all the elements in the periodic table are generated. In a fraction of a second, perhaps 1% of the star's mass is converted into energy. For a period of weeks, the radiant output from the supernova may exceed that of its host galaxy. In a typical galaxy, there is about one supernova per century.

Supernova remnants expand into space at a few percent the speed of light. Heavy elements produced in these eruptions ultimately mix with gases in star-forming regions. Many of the atoms in our bodies were formed in ancient supernovae.

Unless it is destroyed in the explosion, the supernova's core continues to contract. It is too massive to stop at the white dwarf phase. If the original star had a mass about 8–20æ greater than that of the Sun, it will ultimately become a neutron star. Such an object compacts the star's mass into a sphere roughly the size of Manhattan Island. Atomic nuclei and electrons are squeezed into a massive neutron. Neutron stars rotate very rapidly, are often intense radio emitters, and have enormous magnetic fields.

But stars of about 20–50× the Sun's mass keep contracting after the neutron star phase. When they reach a diameter of a few kilometers, they become black holes. Black holes have such a high-mass density that light cannot escape from them. Although stars of solar mass warp space-time near them, the space-time warp produced by black holes is much more significant. It is not impossible that material falling into a black hole may emerge in a distant region of space-time or even another universe.

PAVEL PARENAGO AND HIS DISCONTINUITY

Pavel Petrovich Parenago (1906–1960), was a Soviet-era Russian astronomer. According to the Wikipedia article describing his career, he directed the Department of Stellar Astronomy at Moscow State University and was a Corresponding Member of the Soviet Academy of Sciences. He received the Order of Lenin for his accomplishments. Both an asteroid and a crater on the Moon's far side are named for him.

Being a researcher with unusual intellectual abilities living and working in Soviet Russia during challenging times, Parenago was faced with the problem of how to avoid being sent to a Siberian Gulag, which was the fate of many of his colleagues. He insured his survival by dedicating a book on astronomy to that great genius, Comrade Josef Stalin!

One of his accomplishments was to survey the velocities of nearby stars in their orbits around the center of the Milky Way. As is true of our Sun, most nearby stars circle the galaxy's center at about 220 kilometers per second and require about 225 million years to complete one galactic revolution. Since the demise of the nonfeathered dinosaurs 65 million years ago, the Sun has completed about one quarter of an orbital circuit.

Parenago must have been intrigued as he evaluated the data and discovered what has become known as Parenago's discontinuity. Among his sample of the galactic star population, he found that red, cooler, low-mass stars such as the Sun orbit the galactic center at a slightly higher velocity than the Sun's sisters which are blue, hot, more massive and shorter lived.

When I began to investigate the possibilities of stellar consciousness, I decided to check Parenago's discontinuity for the velocities of main sequence stars in the direction of the Sun's galactic rotation. I located two sources for

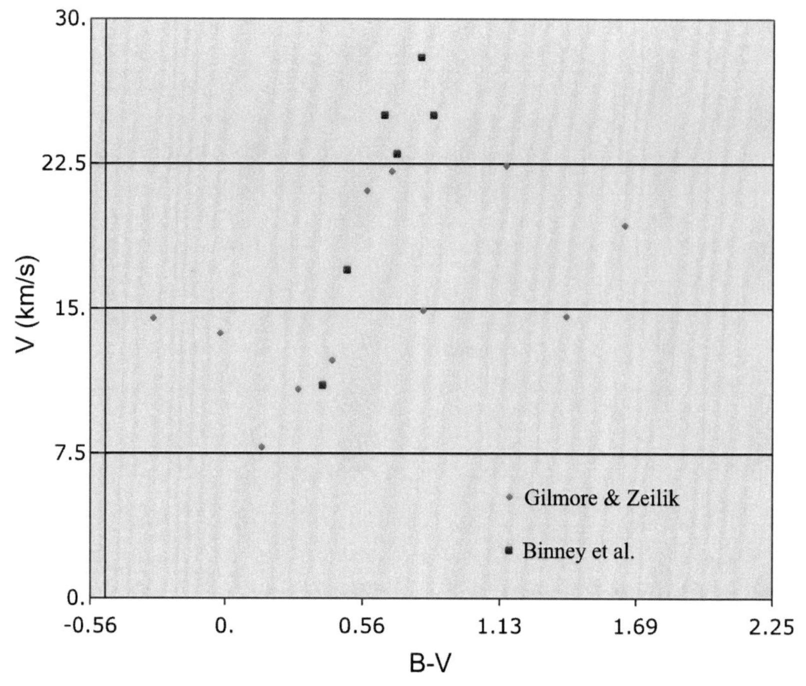

Figure 13.3. Star motion in the direction of the Sun's galactic rotation (V) versus (B-V) color index.

this information. The first was table 19.11.1 in G. F. Gilmore and M. Zeilik, *Allen's Astrophysical Quantities*, 4th edn., a standard reference of astrophysical data. The second was from a 1997 paper by J. J. Binney and colleagues that deals with measurements of the motions of 5610 main sequence stars within 260 light years of the Sun using data obtained from the European Hipparcos satellite.

The combined data from these two sources has been plotted and is presented in Figure 13.3. The velocities in this figure are relative to the local standard of rest. This is a reference frame, described by Gilmore and Zeilik, which is defined from the mean motion of nearby stars around the galactic center. Note from Figure 13.3 that stars with color index (B-V) greater than about 0.5 typically move 20 kilometers per second faster than stars with lower color indices in their orbits around the galactic center.

Table 13.1 presents data that may help in evaluating the significance of this discontinuity. Color indices are tabulated versus spectral class with greater precision than in Figure 13.1. The velocity discontinuity at roughly B-V = 0.5 corresponds to a main sequence stellar spectral class of about F7. Such F7 stars are slightly hotter, bluer, more massive, and shorter lived than the Sun.

As discussed in a classic reference authored by Lawrence Aller and a somewhat more recent paper by G. F. Sitnik and M. Ch. Pande, molecules are

Table 13.1. Main sequence star spectral classes and (B-V) color indices.

Star spectral class (B-V) color index	
B0	−0.30
B5	−0.17
A0	−0.02
A5	0.15
F0	0.30
F5	0.44
G0	0.58
G5	0.68
K0	0.81
K5	1.15
M0	1.40
M5	1.64

rare or nonexistent in the spectra of hot, blue stars. The spectral signatures of molecules such as CO (carbon monoxide) have been detected in solar spectra. The infrared signature of M2 main sequence stars indicates the presence of more complex molecules including TiO (titanium oxide) and ZrO (zirconium oxide).

CONCLUSIONS: WHAT IS THE CAUSE OF THE DISCONTINUITY?

It is certainly interesting that Parenago's discontinuity occurs almost exactly at the place in the stellar continuum where molecular signatures begin to appear in stellar spectra. Binney and colleagues speculated that the cause for this might be that gravitational scattering increases with star age. But stars are only close enough together for this to occur in their birth nebula. Moreover, as discussed by Chaisson and McMillan and many other sources, stellar birth nebula typically disperse within a few hundred million years. If this were the cause, the discontinuity in velocity would occur at a negative value of (B-V). Another cause has been suggested: galactic density waves. However, as will be discussed in Chapter 18, early observational evidence does not support this hypothesis. As demonstrated in Chapter 14, Parenago's discontinuity is supporting observational evidence for the volitional star hypothesis.

FURTHER READING

There are many historical treatments of the development of the art of stellar classification. A good one is O. Struve and V. Zebergs, *Astronomy of the 20th Century* (Macmillan, NY, 1962). Any college astronomy text will yield more information on the art of stellar classification, as will many online sources. My source is E. Chaisson and S. McMillan, *Astronomy Today*, 6th edn. (Pearson Addison-Wesley, San Francisco, CA, 2008). This reference is also my source for the velocity and period of the Sun's orbit around the center of the Milky Way.

Information on Pavel Petrovich Parenago is available online (*http://en.wikipedia.org/wiki/Pavel_Petrovich_Parenago*). Parenago's ploy of dedicating one of his books to that great humanitarian hero, Stalin, is described online by Google books at *http://books.google.com/books?id=eq7TfxZOzSEC&pg=PA260&lpg=PA260&dq=parenago+and+Stalin&source=bl&ots=Sg_KHOtY11&sig=TRMndHwEZd3Ax39q_4LCCWynQJU&hl=en&sa=X&ei=Oh7ZU8ybPNbLsASwmIGwCQ&ved=0CCwQ6AEwAg#v=onepage&q=parenago%20and%20Stalin&f=false*

Some of the data on main sequence stellar kinematics are from G. F. Gilmore and M. Zeilik, "Stellar populations and the solar neighborhood," chapter 19 in *Allen's Astrophysical Quantities*, 4th edn. (ed. A. C. Cox, Springer-Verlag, NY, 2000). Additional data are from J. J. Binney, W. Dehnen, N. Houk, C. A. Murray, and M. J. Preston, "Kinematics of main sequence stars from Hipparcos data," in *Proceedings ESA Symposium Hipparcos Venice '97, ESA SP-402, Venice, Italy, May 13–16, 1997*, pp. 473–477 (European Space Agency, Paris, 1997).

The (B-V) color indices of main sequence stars are from table 15.7 of J. S. Drilling and A. U. Landolt, "Normal stars," chapter 15 in *Allen's Astrophysical Quantities*, 4th edn. (ed. A. C. Cox, Springer-Verlag, NY, 2000).

A classic reference discussing molecular spectral signatures versus main sequence star spectral class is L. W. Aller, "Interpretation of normal stellar spectra," chapter 5 in *Stellar Atmospheres, Vol. 6: Stars and Stellar Atmospheres* (ed. J. L. Greenstein, University of Chicago Press, Chicago, IL, 1960). A slightly more recent paper on this topic is G. F. Sitnik and M. Ch. Pande, "Two decay processes for CO molecules in the solar photosphere," *Soviet Astronomy*, **11**, 588–591 (1968).

CHAPTER 14

The volitional star hypothesis

With wonderful deathless ditties
We build up the world's great cities,
And out of a fabulous story
We fashion an empire's glory:
One man with a dream, at pleasure,
Shall go forth and conquer a crown;
And three with a new song's measure
Can trample an empire down.

Arthur O'Shaughnessy, *Ode*

I do not expect that the hypothesis proposed here—that stars are volitional—will trample any empires. Nevertheless, it is not beyond the realm of reason that this dream of Olaf Stapledon, presumably derived at pleasure, can help alter the existing paradigm of the role of consciousness in the Universe. The hypothesis will be presented and examined as a series of axioms.

AXIOM 1. A UNIVERSAL FIELD OF PROTO-CONSCIOUSNESS EXISTS

It is assumed, in accordance with many interpretations of quantum mechanics and the previously discussed theory of Roger Penrose, that consciousness at all levels results from the interaction of physical matter with a universal field of proto-consciousness. This is not pantheism, since this field may have arisen randomly in one of a gazillion universes created in a multiverse, rather than by the conscious action of a divine designer. It should probably be classified as panpsychism (the philosophical doctrine that mind at some level is a universal feature of everything).

This philosophical approach contradicts the conception of consciousness as an epiphenomenon (something that emerges from matter as neuronal structures become more complicated). However, there is no disagreement with the consideration that molecular structures are less conscious than biological cells with microtubules, simple neuronal structures are more conscious than microtubules, and animal consciousness increases with neuronal complexity.

If correct, panpsychism solves a paradox of many quantum mechanical interpretations. If a conscious observer is necessary to collapse a wave function to convert probabilities to events in the real world, a universal field of proto-consciousness created with the Universe could perform this function in the eons before the evolution of organic brains.

AXIOM 2. FLUCTUATIONS IN THE UNIVERSAL VACUUM ARE THE SOURCE OF THIS FIELD

In modern cosmology, the basic creative act in the Big Bang, the origin of the Universe, is stabilization of a fluctuation in the universal vacuum or quantum foam that underlies physical creation (Figures 14.1 and 14.2). As demonstrated theoretically by Casimir and since validated in replicable experiments, the pressure produced by vacuum fluctuations is a significant contributor to the agency binding atoms within molecules.

This axiom implies that, in a sense, all molecules are conscious. Although strict materialistic reductionists may object to this proposition, I contend that

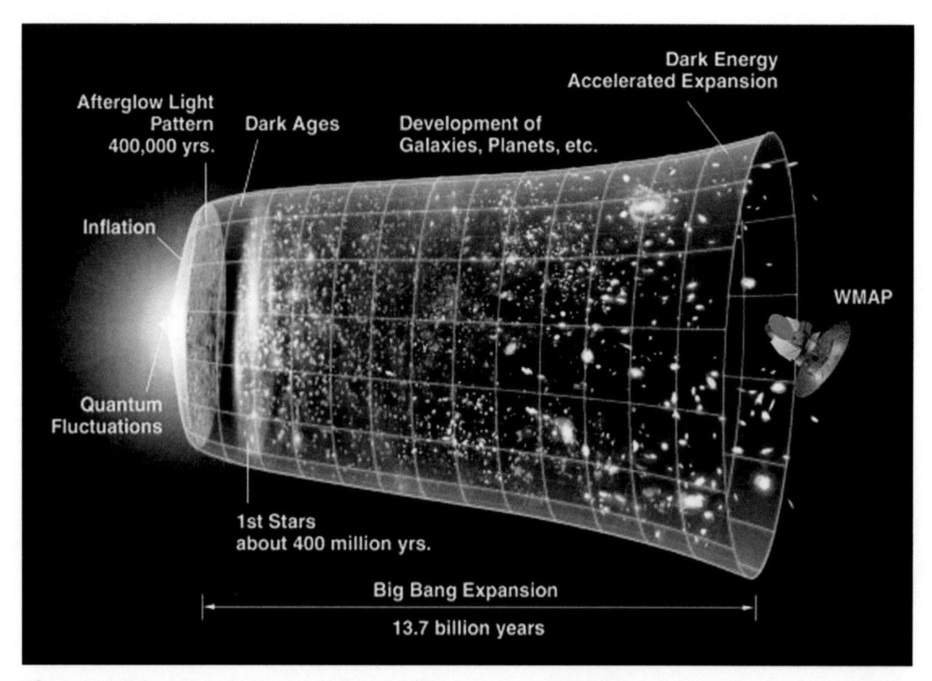

Figure 14.1. Microwave-sensitive satellites such as WMAP can view the Universe a few hundred thousand years after the Big Bang (courtesy NASA).

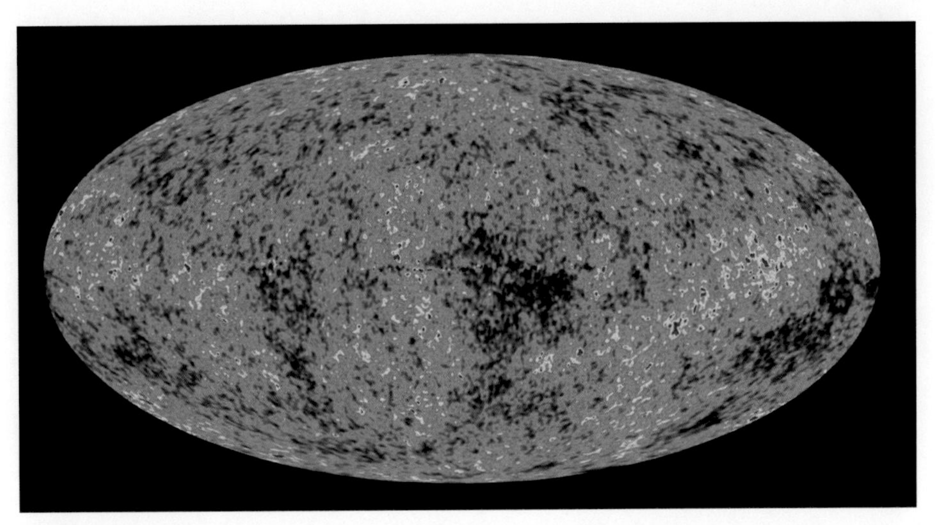

Figure 14.2. Microwave image of the Universe at about 370,000 years old (courtesy NASA).

not including vacuum–fluctuation pressure in their definition of matter has been obsolete for decades.

AXIOM 3. STARS WITH SIGNIFICANT QUANTITIES OF MOLECULES ARE CONSCIOUS

Stars are immense objects, at least from the viewpoint of mortal humans. It is most interesting that those stars with simple and complex molecules in and below their photospheres move differently than those too hot to possess molecules. Stellar consciousness may be regarded as a possible explanation for Parenago's discontinuity.

Nothing is said here about the level of stellar consciousness. As some have suggested, stellar consciousness may be at the level of a classical god or no higher than that of a slime mold amoeba.

AXIOM 4. VOLITION IS AN ASPECT OF STELLAR CONSCIOUSNESS THAT WE CAN OBSERVE

Mainstream explanations for the motions of stars in the outer fringes of spiral galaxies such as our Milky Way do not succeed. To explain these motions, dark matter in the form of massive astrophysical compact halo objects (MACHOS) and weakly interacting massive particles (WIMPS) have been invoked. However, there is insufficient observed disruption to globular clusters in galactic haloes and too few gravity lens events to support the MACHO hypothesis. The search for massive particles that only react with visible matter in a gravitational manner, have no electric charge, and avoid the inner regions of galaxies while constituting about 65% of the Universe's mass has thus far failed. Attempts to modify Newtonian dynamics (MONDS) to account for anomalistic stellar motions also fail since different modifications are necessary at galactic and cosmological distances.

As an alternative to these proposals, we suggest that Olaf Stapledon's concept that volitional stars are equipped with a herding instinct and follow their trajectories out of choice should be further examined. If confirmed by future observations, the hypothesis of volitional stars will require substantial modification to existing scientific paradigms. But these metaphysical constructs have changed before in the history of human thought. There is no reason to suspect that they will not change again.

AXIOM 5. WE CAN DEFINE, QUANTIFY, AND PERHAPS OBSERVE METHODS STARS MIGHT USE TO ALTER THEIR MOTIONS

It is possible to define possible methods that a volitional star might use to alter its gravitationally defined orbit around the center of its galaxy. Because observations can be used to verify or falsify the volitional star hypothesis, it should be considered a scientific concept rather than a metaphysical construct. Methods that a volitional star could use to alter its galactic trajectory are further considered in Chapter 15. The observational consequences of such kinematics modifications are discussed in Chapters 15, 17–20, and 23.

AXIOM 6. PERHAPS GALAXIES AND LARGER STRUCTURES ARE SELF-ORGANIZING

Chapters 22 and 23 consider speculation about and evidence for some form of self-organization operating in structures larger than stars, such as galaxies. Perhaps the Universe as a whole has a tendency for self-organization built into its structure. Stellar volition in a galaxy might be analogous to the tendency of cells in a living organism to self-organize into organs.

IMPLICATIONS: WHAT IF PANSYCHISM IS OBSERVATIONALLY CONFIRMED

The most radical aspect of this book is the proposal that a theory of consciousness could be confirmed or falsified by astrophysical observations. It is reasonable to wonder what the consequences would be for human civilization if the study of consciousness emerges from philosophy and metaphysics to become a subdivision of science.

The first question I ask myself is what will be the effect on world religions? Many scientists have adopted, at least in public, an atheistic stance. This may be because of the suffering imposed by religious individuals and groups over the millennia on free thinkers such as Hypatia, Bruno, Galileo, and Spinoza. Religion certainly has a downside. Nevertheless, it can be argued that the threat of eternal punishment and the lure of eternal bliss has also served to control the baser tendencies of some people. Moreover, should we collectively ignore all aspects of revealed knowledge, the basic moral and legal codes of civilization will lose their legitimacy.

An observational confirmation of panpsychism might be welcomed by Eastern religions, particularly some interpretations of Buddhism. According to some Buddhist thinkers, there is no personal god. Everything emerges from the void and ultimately returns to it. Although modern Hinduism has a multitude of gods and goddesses as did classical pantheons, earlier Vedic sources might consider these to be symbols of universal attributes. A panpsychic interpretation of consciousness as an intrinsic property of the Universe might be acceptable.

The Western monotheistic religions—Judiasm, Christianity, and Mohammedism—might on the surface have more difficulty. Could universal consciousness be congruent with concepts of a divine creator and human souls?

My rather tentative answer to this conjecture is "perhaps." What panpsychism states is that consciousness or the potential for consciousness is built into the structure of the Universe. This Universe of ours could be one of a multitude that evolved at random with just the right properties for our emergence, including a field of proto-consciousness. Or it could be the single production of a creative entity, with the proto-consciousness field being divine "breath" animating the "dust" of physical creation. Alternatively, as in Stapledon's *Star Maker*, our conscious Universe could be one of a huge or infinite number of series or parallel universes created by a divine agent.

Regarding the concept of the human soul and its fate after the termination of physical life, it seems safe to say that science has little to add to theology. Unless consciousness researchers can someday perform experiments or gather repeatable observations in the post-death realm, religious faith is not threatened by the concepts of universal proto-consciousness.

Panpsychism may be new to the halls of science, but is certainly no stranger to philosophers. Its 2700-year pedigree begins with the pantheism proposed by Thales of Miletus and later by Plato, and continues with the philosophical contributions of such luminaries as Baruch Spinoza, Gottfried Wilhelm Leibniz, and Arthur Spinoza. Other contributors to its development were the pioneering American psychologist William James and Pierre Teilhard de Chardin, a Jesuit priest who contributed to 20th century paleontology.

It has been discussed in not unfriendly terms in sources as diverse as the *Scientific American Mind* and the *Catholic Encyclopedia*. The prospect that we might be able to test some of its precepts scientifically is exciting and intriguing.

FURTHER READING (AND VIEWING)

There are many online and published sources dealing with the doctrines of panpsychism. A good one is an article in a peer-reviewed online publication, the *Internet Encyclopedia of Philosophy* (*http://www.iep.utm.edu/panpsych/*).

For a theological approach to this subject, you might wish to check out the *Catholic Encyclopedia* (*http://www.newadvent.org/cathen/11446a.htm*). I wonder if Pierre Teilhard de Chardin's interest in this topic influenced the authors of this article.

Another approach to this subject is C. Koch, "Is consciousness universal?" *Scientific American Mind*, **25**(1), December 19, 2013 (*http://www.scientificamerican.com/article/is-consciousness-universal/*). Christof Koch is the chief scientific officer of the Allen Institute for Brain Science in Seattle and serves on *Scientific American Mind*'s board of advisors.

Many contemporary philosophers are considering and discussing panpsychism. Notable among them is David J. Chalmers, who is affiliated to both the Australian National University and New York University. At least one of his lectures on the topic, "Panpsychism and Panprotopsychism," is available (*http://consc.net/papers/panpsychism.pdf*). His term in the essay's title, "panprotopsychism," refers in a certain sense to the universal proto-consciousness field proposed by Roger Penrose and other quantum physicists.

In David Chalmers' TED talk on consciousness, "How do you explain consciousness," panpsychism is discussed. In this talk, which was published on YouTube on July 14, 2014, Chalmers expresses the opinion that the time is now ripe for science to attempt an explanation of consciousness. Panpychism may be a better explanation for human and animal consciousness than treating this as an emergent phenomenon. You can access this very entertaining and informative talk at *https://www.youtube.com/watch?v=uhRhtFFhNzQ*

CHAPTER 15

Stellar jets and psi

> I know I am deathless,
> I know this orbit of mine cannot be swept by a carpenter's compass,
> I know I shall not pass like a child's curlicue cut with a burnt stick at night.
>
> Walt Whitman, *Song of Myself*, from *Leaves of Grass*

If stars are in some sense volitional and if this is at least a partial solution to the kinematics anomalies posed by dark matter and Parenago's discontinuity, how might a conscious star adjust its galactic orbit? This is the subject of this chapter.

We suggest three methods by which a conscious star could alter its motion. The first is unidirectional emission of photons which, while physically possible, has never been observed. The second involves unidirectional (or unipolar) matter jets from young stars and has been observed, as discussed in our article on this topic in the *Baen Press* online science magazine. The third involves some form of psychokinetic action of the star on itself and is very controversial, as discussed in Chapter 7.

Calculations are limited to stars of solar mass ($\sim 2 \times 10^{30}$ kg). It is assumed that the agency altering a star's galactic trajectory operates during the first 10 billion years or 10% of the star's main sequence lifetime. Moreover, since Parenago's discontinuity indicates unexplained velocity differences between hot and cold stars of about 20 kilometers per second, this will be the required velocity increment.

In each year, there are about 3.16×10^{7} seconds. Therefore, in 1 billion years, there are about 3.16×10^{16} seconds. Since 1 kilometer = 1000 meters, the subject star changes its velocity by 20,000 meters per second in 1 billion years. This corresponds to an acceleration (change in velocity/time) of about 6×10^{-13} meters per square second.

We will have to watch the stars for a long time and greatly increase the precision of our measurements to verify (or falsify) the possibility of such stellar acceleration. During a 100-year human lifetime, this acceleration results in a velocity change of about 0.2 centimeters per second! This acceleration is about 6×10^{-14} of Earth's gravitational acceleration at the surface of the planet.

The force generated by this acceleration is the product of the star's mass and the acceleration. The average or constant force generated by the star is about 1.2×10^{18} newtons. For comparison, the combined thrust at lift-off of the three main liquid rocket engines and two solid strap-on solid rockets of the U.S. space shuttle was less than 3×10^{7} newtons. Billions of space shuttles would have to be launched on a constant basis to equal the force required to change the Sun's velocity by 20 kilometers per second during a 1-billion-year time interval.

Now that the required force has been quantified and put into human terms, attention is turned to how a conscious star might generate this force. The first option considered is unidirectional radiation pressure.

OPTION 1. UNIDIRECTIONAL ELECTROMAGNETIC RADIATION PRESSURE

Electromagnetic photons have no mass. But, surprisingly, they do have linear momentum. In classical physics, the linear momentum of an object is the product of its mass and velocity. The linear momentum of the massless electromagnetic photon is defined in quantum mechanics as h/λ, where h is Planck's constant and λ is the photon's wavelength. In MKS units, Planck's constant is 6.63×10^{-34} joule-seconds.

A yellow photon near the peak of our Sun's spectral energy distribution curve has a wavelength of about 5×10^{-7} meters. This means that every yellow photon emitted by the Sun has a linear momentum of about 1.3×10^{-27} in the MKS system of units.

This is a tiny number! But, the Sun, and other main sequence stars, emit copious numbers of electromagnetic photons. It is in fact the pressure of these photons, as they rise through the stellar interior to ultimately be emitted from the photosphere, that keeps the star from collapsing. The radiation pressure force of these photons, which are the result of thermonuclear processes deep within the star, exactly balances the star's self-gravitation. Without radiation pressure, the star would rapidly collapse.

It is assumed by virtually all astronomers and backed up by centuries of observation, that radiation pressure is omnidirectional. If it were not, the star's gravity would not be uniformly balanced. The Sun (and other stars) would lose their spherical appearance.

But, might it be possible that the balance between gravity and radiation pressure is not always perfect? Might there be unidirectional emission of photon beams from young stars that could account for the velocity change defined above?

This seems most unlikely. One way to express the radiation pressure force of a photon beam is P/c, where P is the radiant power of the beam, in watts and c is the speed of light in vacuum (3×10^{8} m/s). Equating this expression with the required force of 1.2×10^{18} newtons, we find that the required radiant power in the unidirectional photon beam is about 4.6×10^{26} watts. Because this exceeds the radiant power emitted by the present-day Sun (3.9×10^{26} watts), unidirectional electromagnetic radiation pressure does not seem to be a candidate trajectory alteration mechanism for volitional stars.

Nevertheless, I still wonder if the assumption of absolute omnidirectional radiation pressure is entirely correct. After all, at any given time, we observe starlight and sunlight from only one direction. Perhaps this assumption could someday be checked using a constellation of solar observatories, all at the same distance from the Sun but in different orbital planes. Very accurate observations of solar radiant emissions simultaneously and from many locations might be very revealing.

OPTION 2. UNIPOLAR MATTER JETS

All young and mature stars, including the Sun, emit streams of electrically charged subatomic particles into their immediate environment. The velocity of this highly variable solar wind is typically 400–800 kilometers per second. At the orbit of the Earth, the solar wind density varies between 10 and 100 ions per cubic centimeter. In a typical year, the Sun ejects 10^{-13} of its mass in the solar wind.

The Hubble Space Telescope and other modern instruments have observed that young stars often eject bipolar, symmetrical jets (Figure 15.1). When I composed the *JBIS* paper and the *Centauri Dreams* blog piece, I believed that all observed stellar jets were of this nature.

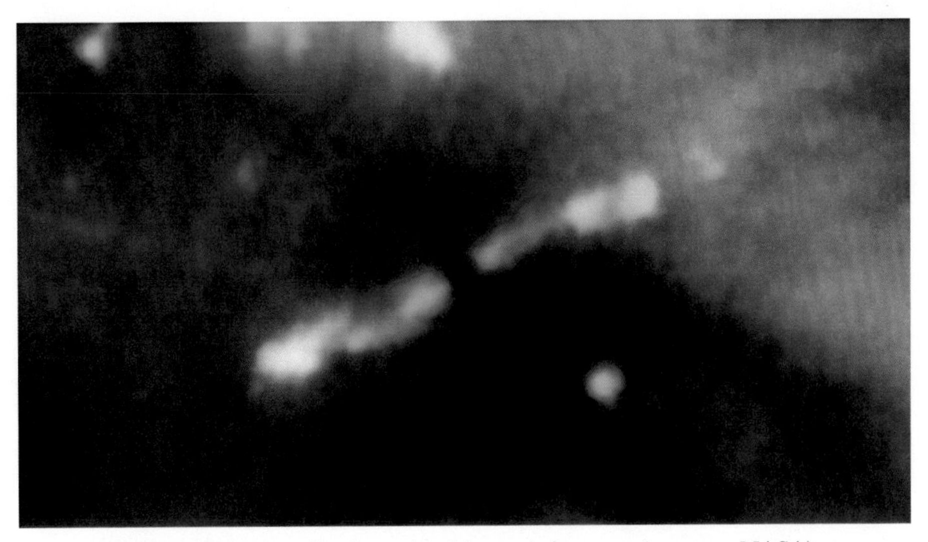

Figure 15.1. A stellar jet emitted by an infant star (courtesy NASA).

For a matter jet to alter the galactic trajectory of a young star, it would have to be asymmetric, with different density and velocity for material emitted from the two poles of the jet. Surprisingly, such jets have been detected with jet velocities varying by a factor of at least 2!

Observations have revealed that much material is ejected in these jets from infant stars. According to a paper by S. R. Cranmer, a typical young star ejects mass at an annual rate 10^{3}–10^{6} that of the Sun.

Jet velocity and density have been measured for a number of young stars. For five infant T Tauri stars, jet velocity has been determined to vary between 80 and 167 kilometers per second, somewhat slower than the typical solar wind. The jet density for these stars is 10^{5}–10^{6} ions per cubic centimeter.

In my *Baen Press* online article cited in this chapter's "Further reading," I argue from the data just presented how a unipolar or asymmetric stellar material jet is a very viable candidate for the means by which an infant star could alter its galactic trajectory. Consider, for example, an infant star of approximately solar mass at its T Tauri stage which ejects an asymmetric jet for a time period of 2 million years, at a jet velocity of 100 kilometers per second and an annual mass ejection rate one million times that of the Sun or 10^{-7} solar masses per year.

The total amount of mass ejected is 0.2 solar masses or 20% of the star's approximate mass. Since the jet velocity is 100 kilometers per second, we can apply the principle of conservation of linear momentum (mass $\times$ velocity) to estimate that the star's galactic velocity changes by about 20 kilometers per second during this time interval. This is done by equating the magnitude of the jet's linear momentum change to the magnitude of the star's linear momentum change.

We can also, from the definition of kinetic energy, estimate the fraction of the present-day Sun's radiant power output which must be applied to energize the matter jet from this infant star. The power (in watts) required is one half the product of the jet's mass emission rate in kilograms per second and the square of the jet velocity in meters per second. Since the star ejects 20% of the Sun's current mass (4×10^{29} kg) in two million years, the mass ejection rate is about 6.3×10^{15} kilograms per second and the jet velocity is 100,000 meters per second. The amount of power required to energize the jet is therefore about 3.2×10^{25} watts. This is about 8% of the present-day radiant output of our Sun.

It should be remembered that, in interpreting this result, our data on bipolar and unipolar stellar jets are very limited. At best, these data represent a snapshot of a long-duration process. As discussed in Chapter 20 on the observational evidence required to validate the volitional star hypothesis, information on average stellar jet alignment relative to the direction of stellar motion and average stellar jet intensity relative to distance from the galactic center is required.

Although unipolar stellar jets are probably the best candidate for the technique a volitional star might use to alter its galactic trajectory, there may be other alternatives. One, as discussed in the next section, is a very controversial (and perhaps nonexistent) weak psychokinetic force.

OPTION 3. A WEAK PSYCHOKINETIC FORCE

As discussed in Chapter 7, a group of physicists during the 1960s–1970s attempted to scientifically evaluate the possibility that the force of human will can operate directly on matter to alter motion at a distance without contact. Although their efforts were stymied when it was demonstrated that their best-scoring psychic was a talented magician, many of the researchers remain convinced that their tentative conclusions based on original screening data are accurate.

Although the existence of psychokinesis remains unproven, it certainly cannot be completely ruled out at this time. This section attempts to evaluate the necessary strength of a weak psychokinetic force generated by a volitional star to alter the velocity of its galactic trajectory by 20 kilometers per second in 1 billion years.

As discussed above, the required force to accelerate a star of solar mass ($\sim 2 \times 10^{30}$ kg) by 20 kilometers/second during a 1-billion-year time interval is 1.2×10^{18} newtons. If a hypothetical psychokinetic force scales with the mass of the entity producing the force, a 100-kilogram human could generate a psychokinetic force of 6×10^{-11} newtons.

To see what these numbers mean in human terms, consider a square piece of hyper-thin aluminum foil with a thickness of 1 micron (10^{-6} m or 0.001 mm). The density of aluminum is 2.7 grams per cubic centimeter. To just levitate a piece of this foil off the ground, the psychokinetic force must exactly compensate for the weight of the foil, which is the product of its mass and the acceleration of gravity at Earth's surface (9.8 m per square second). Since density is the ratio of mass to volume and the volume of the square foil segment is the product of its area and thickness, the maximum foil mass is about 6×10^{-9} grams. The maximum dimension of the square foil is about 0.05 millimeters!

In human terms, the force required is exceedingly small, very much less than that required to bend a fork! Perhaps this could be demonstrated and measured having test subjects concentrate on foil segments placed in a transparent, sealed vacuum chamber. However, as will be discussed below, another possibility has been suggested by someone who commented on my contribution to Paul Gilster's *Centauri Dreams* blog.

The star-generated power that must be applied to modify the star's galactic velocity by this process can also be estimated. This can be done using a classical definition of power (mass × velocity × acceleration) and considering that the star initially circles a galactic center reference frame at a velocity of 200 kilometers per second (2×10^{5} m per second). The star's mass is 2×10^{30} kilograms and its acceleration is about 6.3×10^{-13} meters per square second. The approximate power required is therefore about 2.5×10^{23} watts, about 0.00065 of the present-day Sun's radiant output.

CONCLUSIONS: FUTURE WORK POSSIBILITIES

Researchers interested in further investigation of the methods a volitional star might use to alter its galactic trajectory are encouraged to stay abreast of observations of unipolar stellar jets. Two questions in particular need to be answered. How are these distributed relative to the galactic center? How does average jet velocity vary with distance from the galactic center?

Despite psychokinesis being a controversial subject, there is a need for replicable experiments to be designed and attempted to measure the very weak psychokinetic force that might be built into human consciousness as well as stellar consciousness.

In addition, it is usually assumed that solar and stellar radiant output is omnidirectional. But is this assumption absolutely correct? Future constellations of solar observatories in different solar orbits might be necessary to clarify this issue.

Finally, it should not be assumed that the three modes of stellar self-acceleration suggested here exhaust all possibilities. Creative theoreticians are encouraged to suggest others.

FURTHER READING

Our contribution to the *Baen Press* online science magazine on stellar consciousness is entitled "Star that Wander, Are You Bright?" (*http://www.baen.com/starsconscious.asp*). Art by C Bangs is included in that piece.

The performance characteristics of contemporary rocket engines are available from many sources. I used M. J. L. Turner, *Rocket and Spacecraft Propulsion: Principles, Practices, and New Developments*, 2nd edn. (Springer–Praxis, Chichester, U.K., 2005).

My reference for Planck's constant and the linear momentum of a photon is A. Messiah, *Quantum Mechanics*, Vol. 1 (Wiley, NY, 1976).

The equation used for the radiation pressure force of a unidirectional photon beam is a modification of the equation commonly used in considerations of spacecraft propelled by power beams. See G. L. Matloff, *Deep-Space Probes*, 2nd edn. (Springer–Praxis, Chichester, U.K., 2005).

My source for the Sun's present-day radiant power was E. Chaisson and S. McMillan, *Astronomy Today*, 6th edn. (Pearson Addison-Wesley, San Francisco, CA, 2008).

Solar wind characteristics are from J. B. Kaler, *Astronomy* (HarperCollins, NY, 1994). The annual rate at which the Sun is ejecting mass to the solar wind is from P. D. Noerdlinger, "Solar Mass, the Astronomical Unit, and the Scale of the Solar System" (*http://arxiv.org/abs/0801.2807*).

A good paper on asymmetric stellar material jets is F. Namouni, "On the flaring of jet-sustaining accretion disks," *Astrophysical Journal*, **659**, 1505–1510 (2007). My source on the mass ejection rate from typical infant stars is S. R. Cranmer, "Turbulence-driven polar winds from T Tauri stars energized by magnetospheric accretion," *Astrophysical Journal*, **669**, 316–334 (2008).

The data I used for jet density and velocity from typical T Tauri stars are from A. I. Gomez-de Castro, E. Verdugo, and C. Ferro-Fontan, "Wind/jet formation in T Tauri stars: Theory vs. UV observations," in *Proceedings of 12th Cambridge Workshop on Cool Stars, Stellar Systems, and the Sun, July 30–August 3, 2001*, pp. 40–49 (University of Colorado, Boulder, CO, 2003).

Any university level first-year physics text will reveal that the power required to maintain an object's motion relative to a reference frame is the product of the magnitude of the force on that object and the object's velocity, or object mass × acceleration × velocity. My reference is F. W. Sears, M. W. Zemansky, and H. D. Young, *University Physics Part 1*, 5th edn. (Addison-Wesley, Reading, MA, 1977).

CHAPTER 16

Of publishing and blogging

The people know the salt of the sea
and the strength of the winds
lashing the corners of the earth.
The people take the earth
as a tomb of rest and a cradle of hope.
Who else speaks for the Family of Man?
They are in tune and step
with constellations of universal law.

Carl Sandburg, *The People Will Live On*

For centuries, the people of Earth have passed knowledge down to their descendants and commented on this knowledge using the medium of paper. Today, this has completely changed. Nowadays, alterations to global consciousness move a great deal faster. The emergence of the blogosphere is perhaps the most revolutionary improvement in the rate of information transfer in five centuries. For better or worse, it promises to alter the rate at which new concepts are embedded in human civilization's thought processes.

My only personal interaction with this new medium is my previously discussed entry on stellar consciousness in Paul Gilster's *Centauri Dreams* blog. Before further discussing this experience, I will review the history of the art of human information transfer.

HUMAN INFORMATION TRANSFER: A BRIEF HISTORY

Transfer of information between individual human beings can be thought of as a form of self-transcendence. Using the techniques of information transfer, we can expand our ideas and emotions beyond the confines of our brains and bodies. As these technologies became more sophisticated, it became possible for our mental constructs to escape the limitations imposed by our physical lives.

Many animal species use a form of language. Bees dance to inform their sisters of the location of succulent flowers. Deer grunt to inform other deer of predators in the neighborhood. Bird distress calls warn of prowling house cats.

Early human and prehuman communication must have developed along similar lines.

However, about 1 million years ago, our *Homo erectus* ancestors began to migrate from equatorial Africa on an epic expansion that would ultimately encompass much of Africa and Asia. As they traveled in hunting bands of a hundred or so, the technology of the first human ancestors to walk upright greatly advanced. Fire was used to protect against the cold of temperate climates and possibly for cooking. The pelts of prey animals were used to clothe the body against the elements.

As technologies became more sophisticated, so the earliest proto-human languages must have developed. Words were probably used during hunting, herb gathering, and preparing basic clothing and shelters.

However, there is no way to know for sure. The techniques used to pass knowledge and experience from one generation to the next were likely primitive.

For millennia, progress in spoken language must have been slow though continuous. Around 50,000 years ago, a new level of sophistication was reached. One or more communities of coast-dwelling early humans living in Asia were able to work cooperatively to construct what may have been the first ocean-going craft. Using these crude ships, they island-hopped to colonize the continent of Australia and, ultimately, became the Aboriginal people.

About 15,000 years later, the next breakthrough in human communication occurred. Sometime between 32,500 and 39,000 years ago, pictorial art was invented (Figure 16.1); the Chauvet Cave paintings in France have been dated at 32,000 years ago. Using charcoal and red ochre, an artist or group of Paleolithic (Old Stone Age) artists created small drawings of human and animal figures that have survived on the walls of a cave near Verona in northern Italy. It was arguably through this medium that ideas or emotions were first transmitted to forthcoming generations.

Within a few millennia, cave art had spread throughout Europe. Some of these naturalistic images may have been used in shamanistic rituals to insure the success of the hunt or placate the spirits of animals killed by hunters. Examples of cave paintings of varying age have been found in locations as diverse as Africa, Australia, and the Americas.

The next step in information transfer occurred in the Middle East in the Neolithic (New Stone Age). Around 6000 years ago, many people had given up the nomadic hunter-gatherer way of life and had settled in permanent settlements, which then evolved into the earliest cities.

Life had become more complex. Some of the agricultural communities developed into palace societies as social classes became more diversified. A principal function of the new ruler was to store agricultural produce and redistribute it in times of famine.

This led to a need for some form of record keeping. A little after 4000 BC, some genius derived a partial solution to this problem. Somewhere in Sumeria (modern-day Iraq), the first pictographic script (the world's oldest writing system) was devised. Within a few centuries, cuneiform script impressed on

Figure 16.1. Neolithic African cave painting (courtesy U.S. Library of Congress).

stone tablets (Figure 16.2) was commonplace throughout the Fertile Crescent (a region encompassing the Persian Gulf, southern Iraq, Syria, Lebanon, Jordan, Israel, and northern Egypt), scribes had developed hieroglyphic writing, and Linear A (still undeciphered) was in use in the Aegean. In eastern Asia, many current scripts are directly derived from Bronze Age pictographic systems. These were symbolic scripts, rather than alphabets. For example, when writing about the Sun, a disk symbol would be carved into the stone tablet by the scribe. Thousands of symbols were necessary for the scribe to convey his message. The scribe class would inevitably have been small and in great demand.

During the Bronze Age in Egypt, another genius realized that papyrus scrolls could replace stone tablets. Letters between members of the noble and merchant classes could be transcribed by scribes, sealed with wax imprinted with the sender's stamp, and hand-delivered by couriers. A bit slow by modern standards, but a great improvement over what went before!

Around 1000 BC—as Egypt, Babylonia, and the Aegean recovered from a dark age—the next great advance occurred. Over the next few centuries, the role of the scribes would be greatly diminished as a result of the invention of the first alphabet by someone in the Phoenician city of Tyre, in modern Lebanon. The Hebrew Bible, Greek plays, classical philosophies and histories,

Figure 16.2. Fragment of a 2200–1900 BC Sumerian cuneiform tablet (courtesy U.S. Library of Congress).

and the geometry of Pythagoras and Euclid could all be transmitted more widely using this tool, written on papyrus scrolls and stored in research libraries.

The only major advance between the Classical Period (the 4th and 5th centuries BC) and the European Renaissance was the hand-bound book, which ultimately replaced the scroll in medieval cathedral and mosque libraries. This all changed in around 1450, when the German printer Johannes Gutenberg unveiled a revolutionary invention: the printing press.

No longer would it be necessary to copy a manuscript by hand. No longer would books be so expensive that they rarely, if ever, found their way into the hands of the middle class. It suddenly became possible to set type and transfer it to paper with the aid of a machine. A single Renaissance printing press could produce more than 3000 pages per day. From that time on, literacy and the pursuit of knowledge spread rapidly through the human population.

However, it should not be imagined that Gutenberg worked in a vacuum. During the 11th century, printing from movable type had developed in China. A few centuries later, Korean printers further advanced the art.

It is even possible that experiments with movable type pre-date the Classical Period. At an archeological site in Crete dating from around 1700 BC, a hardened clay disk was discovered in 1908. Called the Phaistos Disk, this artifact seems to have been prepared by pressing hieroglyphic rock or metal seals into the soft clay. The script remains undeciphered, and some suggest that the Phaistos Disk is a modern hoax. However, another explanation is that an invention must be developed at a time that is ready to receive and apply it for it to succeed. The Bronze Age scribes on Minoan Crete may have felt threatened by this radical approach and suppressed the invention.

The next information-transmitting revolution, the one we are living in, began in the mid 19th century with the invention of the electrical telegraph. Applying the new science of electricity and appropriate codes, it became possible to send messages long distances over wires. With the invention of radio, the early 20th century witnessed code and voice messages transmitted long distances without the requirement for cables.

By the middle of the 20th century, television came into wide use. Still and moving visual images could then be transmitted around the world using advanced radio technology. At the same time, facsimile (or Fax) machines facilitated the transmission of scanned printed material over telephone lines.

Now that computing technology and Internet access is available to homes and businesses, the speed of information transfer and the potential audience for this information has become enormous. It is impossible to predict what the ultimate result of this revolution will be, but the rate of innovation shows no sign of slowing. It is not impossible that implanted computer chips directly connected to human brains may break down barriers between individuals and perhaps lead to a form of global brain.

PEER REVIEW VERSUS THE BLOG

There are problems with this information glut. For instance, how do we separate real information from background noise? The producers of scholarly books and journals devised peer review to deal with this problem.

Any manuscript submitted for publication in a scholarly journal is subject to this process. The paper is sent to one or more anonymous readers, generally recognized experts in the field, who read and review it. They may recommend publication, rejection, or suggest appropriate revisions. The process is admittedly not perfect. If the conclusions of the paper contradict the reviewers' own research, it is possible they may reject it. On rare occasions, authors have been known to falsify data in a submitted paper in order to maintain the flow of grant funds. In even rarer cases, papers have been written (and accepted) with meaningless jargon to demonstrate that some reviewers are unable to admit their ignorance.

However, peer review in general has worked rather well. Scholars submitting their work to this process must, unfortunately, accept a time delay of months or longer before their articles are published.

Electronic information transmittal is much faster, almost real time. The late 1980s witnessed a media frenzy surrounding the alleged discovery of table top, room temperature thermonuclear fusion. Because of the vast profits that could result from such a process, renowned research institutions competed to disseminate their unreviewed results as widely and rapidly as possible. The Fax machine was the hero of the moment.

I had a ringside seat to this process since my co-author of the *Starflight Handbook* (Wiley, NY, 1989), Eugene Mallove, was at the time a science writer at MIT. More than one career was severely impacted when it later turned out that cold fusion is extremely hard to replicate. Although it is not impossible that the process releases some energy, almost three decades of research have not resulted in any practical applications.

When I submitted the original manuscript on stellar consciousness (cited in the "Introduction") for publication in *JBIS*, I was not surprised that I had to contend with the comments of four reviewers. This was to be expected because of the controversial nature of the subject. As the editor of that journal suggested, I shortened the article for *JBIS* and contacted Paul Gilster to discuss the possibility of publishing the original manuscript on his *Centauri Dreams* blog.

Paul Gilster is a respected space journalist. I first met him in 2003, when he interviewed experts in theoretical techniques of interstellar travel for his book *Centauri Dreams* (Springer–Copernicus, NY, 2004). After publication of this book, he decided to create a blog of the same name that would present new research in astronomy and astronautics.

I awaited online publication of the blog piece with some trepidation. What was I letting myself in for?

However, perhaps as a result of the article having undergone peer review and as a result of Gilster's excellent job as a blog coordinator, I was very presently surprised. Most of the many responders to the piece were quite serious in their comments. More than a few suggested observational or experimental methods to prove or falsify the conscious star hypothesis. Moreover, this was done almost in real time. If a scholarly blog is well coordinated, there is no reason it cannot serve to rapidly disseminate new concepts to a very large audience.

In Stapledon's *Star Maker*, direct contact with conscious stars is not achieved by organic beings until unified planetary minds or "awakened planets" have evolved. It is not impossible that the first baby steps toward the emergence of a unified terrestrial mind are now being taken through the rapid evolution of the World Wide Web and aspects of it like the blog. It would be nice to come back in a century or two to witness how this revolution has progressed.

FURTHER READING

For information on Paleolithic Italian cave paintings, see M. Balter, "Paintings in Italian cave may be oldest yet," *Science*, **290**, 419–420 (2000). A good source on ancient historical and technological developments is J. Hawkes, *The Atlas of Early Man* (St. Martin's Press, NY, 1976).

A recent paper on ancient art is M. Aubert, A. Brumm, M. Ramil, T. Sutkna, F. W. Saptomo, B. Hakim, M. J. Morwood, G. D. van den Bergh, L. Kinsley, and A. Dosseto, "Pleistocene cave art from Sulawesi, Indonesia," *Nature*, **514**, 223–227 (2014).

Additional information on the technology and history of the printing press is available online (*http://en.wikipedia.org/wiki/Printing_press*). My source for the Phaistos Disk is R. Castleden, *Minoans: Life in Bronze Age Crete* (Routledge, London, 1993). Another source is the Wikipedia entry (*http://en.wikipedia.org/wiki/Phaistos_Disc*).

My source for the history of telegraphy is Wikipedia (*http://en.wikipedia.org/wiki/Telegraphy*). Information on the history of Fax technology is available online (*http://en.wikipedia.org/wiki/Fax*).

CHAPTER 17

Spiral arms: An alternative hypothesis

Elms in thunder,
  the lights in the sky are stars—
We think they do not see,
  we think also
The trees do not know nor the leaves of the grasses hear us

Archibald MacLeish, *Epistle to Be Left in the Earth*

The role of science is not to blindly accept new concepts, no matter how well they are crafted and presented. The development of new explanations to explain physical phenomena constitutes the deductive aspect of physics. Deduction must be balanced and checked by induction (verifying theoretical predictions by observation and experimentation).

Sometime after I had participated in Paul Gilster's *Centauri Dreams* blog, I delved a bit deeper into Parenago's discontinuity. As discussed in Chapter 13, I realized that the one alternative explanation to volitional stars that I was aware of, that of Binney et al., could not be correct. I was surprised to learn that another alternative theoretical concept had been proposed.

MATTER–DENSITY VARIATIONS IN GALACTIC SPACE

This hypothesis has to do with variations in the density of galactic regions. When a high-density region of a spiral galaxy (such as our Milky Way) sweeps through a star field, it might slightly speed up the velocity of lower mass stars in their circuits around the galactic center. This concept has been named the "spiral arm density waves hypothesis."

Figure 17.1 presents a face-on view of a spiral galaxy. Our Sun is about halfway out from the center to the rim, between two high-density spiral arms. Our Milky Way has a diameter of about 100,000 light years and our Sun circles the galaxy's center once every 250 million years. There are a few hundred billion stars in a typical spiral galaxy.

Figure 17.1. The Whirlpool spiral galaxy M51 (courtesy NASA).

The galaxy shown in Figure 17.1 is Messier 51 (M51), the Whirlpool galaxy. This image is from the NASA Hubble Space Telescope and is in visible light. The orientation of the spiral structure indicates that this galaxy rotates in a counterclockwise direction around the bright central hub.

In this image the spiral arms are clearly visible. Most of the material in the interstellar medium is hydrogen and helium, with a smattering of more massive elements. There is more interstellar dust in the star-forming regions of spiral arms than in the intercloud medium, such as our Sun's galactic location.

The density of the intercloud medium is much lower than that of the spiral arms. Near the Sun, there is about 1 hydrogen atom in every 10 cubic centimeters (10^5 hydrogen atoms per cubic meter). The density of star-forming regions in the spiral arms is approximately 1000 hydrogen atoms per cubic centimeter (10^9 hydrogen atoms per cubic meter).

The mass of a hydrogen atom is 1.67×10^{-27} kilograms. One light year is very close to 10^{13} kilometers or 10^{16} meters. A cubic light year is about 10^{48} cubic meters.

A cube one light year on each side contains a volume of about 10^{48} cubic meters. Therefore, a cubic light year of intercloud space contains about 10^{53} hydrogen atoms. This is equivalent to an approximate mass of 1.7×10^{26}

kilograms. This is orders of magnitude less than the Sun's mass of about 2×10^{30} kilograms and is only 30× the mass of the Earth.

Because the hydrogen atom density in a typical interstellar cloud is about 10^4 times greater than that of the intercloud medium, a cubic light year of interstellar cloud contains about one solar mass of material. As discussed below, this density estimate for typical interstellar clouds might be an overestimate.

HOW HIGH-DENSITY REGIONS MIGHT AFFECT THE VELOCITY OF LOW-MASS STARS

But, how might regions of high matter density affect the motion of low-mass stars? Consider the spiral arms of a galaxy, which contain the high-density

Figure 17.2. The Pillars of Creation in the Eagle nebula, M16 (courtesy NASA).

regions, moving through a star field at a higher velocity than that of encountered stars.

The faster moving, high-density gas would tend to attract encountered stars and increase their galactic velocities. Low-mass, cooler, redder stars would be affected more than high-mass, hotter, bluer stars. This might be a possible explanation of Parenago's discontinuity.

But, as further discussed in Chapter 18, there are problems with the spiral arm density waves hypothesis. One is the size and structure of typical high-density clouds. Figure 17.2 presents the famous Hubble Space Telescope image of the Pillars of Creation in the Eagle nebula (M16).

This star-forming region is at an approximate distance of 6000 light years. Its average diameter is about 14 light years. As discussed by J. J. Binney and colleagues (see Chapter 13), data from the European Hipparcos astronomy satellite reveal that Parenago's discontinuity applies to 5610 main sequence stars as far as 260 light years from the Sun.

Because most interstellar clouds are compact and widely separated, it seems unlikely that this kinematics anomaly could be explained by the spiral arm density waves hypothesis.

More significantly, however, an observational test of the spiral arm density waves hypothesis has been performed. Preliminary results, as reviewed in Chapter 18, do not support it. '

FURTHER READING

A paper presenting observational data confirming Parenago's discontinuity and suggesting an explanation based upon star boil-off from the birth nebula is J. J. Binney, W. Dehnen, N. Houk, C. A. Murray, and M. J. Preston, "Kinematics of main sequence stars from Hipparcos data," in *Proceedings ESA Symposium Hipparcos Venice '97, ESA SP-402, Venice, Italy, May 13–16, 1997*, pp. 473–477 (European Space Agency, Paris, 1997). Chapter 13 discussed why this explanation is probably not correct.

Two papers proposing the spiral arm density waves hypothesis for Parenago's discontinuity are: J. Binney, "Secular evolution of the galactic disk," *Galaxy Disks and Disk Galaxies*, pp. 63–70 (ed. F. Bertoli and G. Coyne, ASP Conference Series Vol. 230, Astronomical Society of the Pacific, San Francisco, CA, 2001); and R. S. DeSimone, X. Wu, and S. Tremaine, "The stellar velocity distribution in the stellar neighborhood," *Monthly Notices of the Royal Astronomical Society*, **350**, 627–643 (2004).

As in previous chapters, astronomical conversions were taken from E. Chaisson and S. McMillan, *Astronomy Today*, 6th edn. (Pearson Addison-Wesley, San Francisco, CA, 2008).

According to Chaisson and McMillan, the most compact interstellar clouds have densities as high as 10^6 hydrogen atoms per cubic centimeter. Their table 18.1 shows typical interstellar clouds have matter densities between 80 and 120 atoms per cubic meter. The value used here of 1000 hydrogen atoms per cubic centimeter comes from a classic paper on interstellar travel using interstellar matter as fuel: R. Bussard, "Galactic matter and interstellar spaceflight," *Astronautica Acta*, **6**, 179–194 (1960).

Chaisson and McMillan also cite measurements of matter density in the local intercloud medium. In some regions, this may be as low as 5000 atoms per cubic meter. Because of this density, the intercloud medium is more transparent to ultraviolet light than are typical interstellar clouds. Therefore, the ionization level of the intercloud medium will typically be higher than that of the clouds.

The Hipparcos satellite stellar motion data cited in the text is from J. J. Binney, W. Dehnen, N. Houk, C. A. Murray, and M. J. Preston, "Kinematics of main sequence stars from Hipparcos data," in *Proceedings ESA Symposium Hipparcos Venice '97, ESA SP-402, Venice, Italy, May 13–16, 1997*, pp. 473–477 (European Space Agency, Paris, 1997).

CHAPTER 18

Spiral arms: Not supported by observation

Here is the test of wisdom,
Wisdom is not finally tested in schools,
Wisdom cannot be pass'd from one having it to one not having it,
Wisdom is of the soul, is not susceptible of proof, is its own proof,
Applies to all stages and objects and qualities and is content,
In the certainty of the reality and immortality of things and the excellence of things;
Something there is in the float of the sight of things that provokes it out of the soul.

Walt Whitman, *Song of the Open Road*, from *Leaves of Grass*

For a scientific hypothesis to advance to the stage of becoming a successful theory, it must pass two tests. First, it must be reasonable. Second, it must be verifiable by all observers or experimenters, whether they are true believers or skeptics. In this chapter we investigate whether the spiral arm hypothesis shows promise to be accepted as scientific true wisdom.

Our first test of spiral arms involves the galactic prevalence of comparatively dense clouds or open clusters. These nebulae are, in most cases, the nurseries from which young stars emerge. Most of these objects are located in the spiral arms of the Milky Way.

Astronomers first began to chart these fuzzy, cloud-like objects in the 18th century with their early refracting telescopes. Their goal was, in many cases, to locate and describe objects that might be mistaken for comets. Comet hunting was all the rage at that time.

THE CONTRIBUTION OF CHARLES MESSIER

One of the significant contributors to this sky survey was the French astronomer Charles Messier (1730–1817). As well as observing and cataloging many nebulae, Messier discovered 13 comets between 1760 and 1798. The

Messier catalog contains 104 deep-sky objects. They are not all open clusters. Some, like the Andromeda nebula (Messier 31 or M31), are galaxies external to the Milky Way. Others are compact globular clusters of older stars in our galaxy's halo, such as M13 in Hercules or supernova remnants such as the Crab Nebula (M1).

Table 18.1 presents Messier's open clusters. Celestial coordinates are given in terms of right ascension (R.A.), which is the celestial equivalent of longitude, and declination (Dec.), which is the celestial equivalent of latitude. Modern estimates of distance and approximate dimension are in light years. Common names are listed for some of these objects.

Note that these objects tend to be rather small by celestial standards—typically, 10–30 light years in approximate dimension. Moreover, note that they are widely separated in distance and location on the celestial sphere.

It is certainly not impossible that the Sun and Solar System passed through an open cluster during their 4.7-billion-year history. However, it seems very unlikely that stars out to 260 light years from the Sun passed through the same object, since none of the nebulae tabulated above are larger than 72 light years in size. As discussed in Chapter 13, Parenago's discontinuity is observed for stars at distances up to 260 light years from the Sun. The spiral arm hypothesis cannot be said to have passed this observational test.

The reader should, of course, be skeptical regarding my tentative rejection of the spiral arm density waves hypothesis. After all, I am proposing here and elsewhere a radically different interpretation of Parenago's discontinuity.

Therefore, it would be very nice to have direct observational evidence refuting (or supporting) the predictions of spiral arms. Happily, an observational test of this hypothesis was reported in 2011.

AN OBSERVATIONAL TEST OF SPIRAL ARMS

To understand the observational test of this hypothesis, it is necessary to first discuss the orientation of spiral galaxies in the deep sky. Some will be face on, or top down like the Whirlpool galaxy (M51) shown in Figure 17.1 or M101 shown in Figure 12.2. Some will be edge-on, like the Hubble Space Telescope image of M104, the Sombrero galaxy, presented in Figure 18.1.

Most of the ~100 billion spiral galaxies in the observable Universe are intermediate between these extremes. An example is M66. A Hubble Space Telescope image of this galaxy is presented as Figure 18.2.

One way to test the spiral arms hypothesis is to look for color variations across the spiral arms of external galaxies. A basic tool in this observational testing is the astrospectrograph. This device, usually situated at the prime focus or eyepiece of a large telescope, uses a prism or diffraction grating to split the white light received from a distant celestial object into its constituent colors.

Table 18.1. Messier's open galactic clusters.

M number	*R.A.* (hr, min)	*Dec.* (deg)	*Distance* (lyr)	*Size* (lyr)	*Object name(s)*
6	17 37	S32.2	1300	21	
7	17 51	S34.8	800	22	
8	18 00	S24.4	4850	80	Lagoon
11	18 48	S6.3	5500	18	Wild Duck
16	18 16	S13.8	5870	14	Eagle
18	18 28	S17.2	4900	10	
17	18 18	S16.2	5870	72	Omega
20	17 59	S23	2300	30	Trifid
21	18 2	S22.5	4250	17	
23	17 54	S19	2150	25	
24	18 16	S18.4	16000	19	
25	18 29	S19.3	2060	20	
26	18 43	S9.5	4900	26	
29	20 22	N38.4	4000	11	
35	6 6	N24.3	2850	31	
36	5 32	N34.1	3700	21	
37	5 49	N32.5	3600	30	
39	21 30	N48.2	825	7	
41	6 45	S20.7	1600	24	
42	5 33	S5.4	1500	28	Orion Nebula
43	5 33	S5.3	1500	1	De Mairan's Nebula
44	8 37	N20	525	17	
45	3 44	N24	410	28	Pleiades
46	7 40	S14.7	3200	42	
47	7 34	S14.4	1750	27	
48	8 11	S5.6	1500	27	
50	7 01	S8.3	2950	15	
52	23 22	N61.3	3000	16	
67	8 48	N12	2700	14	
78	5 44	N00.0	1630	3	
93	7 42	S23.7	3600	26	
103	1 30	N60.5	8500	16	

Figure 18.1. Hubble Space Telescope image of M104, the Sombrero galaxy (courtesy NASA).

Figure 18.2. Hubble Space Telescope image of spiral galaxy M66 (courtesy NASA).

If we are observing a face-on or top-down galaxy, the spectral test of spiral arms will be to compare the spectral composition of light received from the stars near the leading edge and trailing edge of a spiral arm.

Assuming that the spiral arm density waves hypothesis is the correct explanation for Parenago's discontinuity, the high-density spiral arms should drag along lower mass red stars as they pass through a star field and leave more massive blue stars behind. This means that the trailing edge of a spiral arm should be a bit bluer than the leading edge.

However, if we observe a galaxy with an intermediate or edge-on orientation, Doppler shift can be applied. According to this principle, which is the one used to explain the expansion of the Universe, spectral lines from the light emitted by objects moving toward us will be shifted toward the blue range of the electromagnetic spectrum. Conversely, spectral lines from light emitted by objects moving away from us will be shifted toward the red. Since spiral arm density waves should speed up less massive red stars and leave more massive bluer stars less affected, Doppler spectral shifts from light emitted by the leading and trailing edges of external galaxy spiral arms should be slightly different.

There are other observational indicators of the effects and durations of spiral arms in external galaxies. Some of these have been applied in an extensive study of 12 nearby spiral galaxies, published in 2011 by a term led by Kelly Foyle of the Max Planck Institute in Heidelberg (Germany).

Foyle and her colleagues were principally interested in the long-term stability of structures in a galaxy's spiral arms. Although the duration of spiral arm structures would affect the duration of density waves, Foyle and her colleagues were primarily interested in star formation rates within spiral arms.

A major result of their observational study, surprising to many who expected long-term stability, is that spiral arm structures are transient rather than long lived. But, what exactly is the definition of "transient" from the point of view of a galaxy? After all, these objects are more than 10 billion years old. Each galactic rotation takes 100 million years or more. The life expectancy of transient spiral arms is a subject of contemporary research. In a very recent paper, Richard N. Henriksen of Queen's University in Kingston (Ontario, Canada) calculates that spiral structures in galaxies last for about 31 million years.

To understand what this means for observations supporting Parenago's discontinuity for stars as far as 260 light years from the Sun, assume that a density wave with a velocity of 20 kilometers per second relative to the average velocity of nearby stars enters a star field with a diameter of 520 light years.

A velocity of 20 kilometers per second is equivalent to about 4 AU per year. Since 1 light year is equivalent to about 60,000 AU, our density wave requires about 15,000 years to traverse 1 light year. It crosses the 520 lyr diameter sphere in about 6.3 million years. Since this is about 6.3 million years of the total predicted life of a spiral structure supporting a density wave, it seems unlikely that stars across a 520 lyr diameter sphere would be similarly affected by this hypothetical density wave.

So, it seems that the hypothesis of a spiral arm density wave being the cause of Parenago's discontinuity observed in the motion of stars as distant as 260 light years from the Sun is probably not correct. To solve the riddle posed by Parenago's discontinuity conclusively, it will be necessary to study the kinematics of more distant stars. This is one of the goals of the recently launched Gaia space observatory. This spacecraft is the subject of Chapter 19.

FURTHER READING

There are a number of catalogs of Messier objects that give position, type, approximate size, and distance. I used K. G. Jones, *Messier's Nebulae and Star Clusters* (American Elsevier, NY, 1969) in the preparation of Table 18.1.

The observational test of the spiral arms hypothesis reported in 2011, which is available both in print and online, is K. Foyle, H.-W. Rix, C. Dobbs, A. Leroy, and F. Walter, "Observational evidence against long-lived spiral arms in galaxies," *Astrophysical Journal*, **735**(2), Article ID = 101 (2011), *arXiv:1105.5141* [astro-ph.CO].

A very recent paper estimating the duration of galaxy spiral structures is R. N. Henriksen, "Transient Spiral Arms in Isothermal Stellar Structures." This paper is in an online archive of physics papers maintained by Cornell University (*http://arxiv.org/pdf/1207.5430v1.pdf*).

CHAPTER 19

A telescope named Gaia

> The poet's eye, in a fine frenzy rolling, doth glance from heaven to Earth, from Earth to heaven; and as imagination bodies forth the forms of things unknown, the poet's pen turns them to shape, and gives to airy nothing a local habitation and a name.
>
> William Shakespeare, *A Midsummer Night's Dream*

In the early 21st century, the role of the "poet's eye" is played to a certain extent by space telescopes. Launched from Earth, these highly crafted mirrors plumb the depths of the heavens. What they see is given shape and name by the mortal terrestrials who manage them. It is not impossible that they will discover new habitations for consciousness, that most tenuous substance in the Universe.

One of the observational requirements to further validate the volitional star hypothesis and perhaps develop it into a full-fledged astrophysical theory is accurate knowledge of stellar motion beyond our local region of the Milky Way. I realize that, in human terms, a distance of 260 light years (about 2.6×10^{15} km) is enormous. But, in terms of the Milky Way, it is small, about 0.5% of our galaxy's radius. The star motion knowledge base is also pitifully small. Accurate stellar motions are known for some thousands of stars. However, our galaxy contains a few hundred billion stars.

A NEW SPACE TELESCOPE

All this will change in the next few years. The European Space Agency (ESA) has launched a space telescope named Gaia. Launched in December 2013 by a Soyuz-Fregat rocket (Figure 19.1), Gaia is currently on station at the Earth–Sun Lagrange 2 (L2) point. With its onboard telescopes, photometers, and spectroscope, it is hoped that this spacecraft will accurately map the positions and motions of about 1 billion stars in the Milky Way galaxy during its 5-year operational mission. The operational phase commenced in July 2014. According to Wikipedia, Gaia is suffering from a modest, stray light problem that may affect observations of the faintest of the billion stars to be studied.

Figure 19.1. The launch of Gaia (courtesy ESA).

Gaia's position in space is at a gravitationally stable location on the Earth–Sun line, about 1.5 million kilometers on the antisunward side of the Earth. The advantages of this location are its stability (relatively little maneuvering fuel is required to maintain position) and the fact that it can use Earth as a partial sunshade to reduce the amount of stray light.

In late July 2014, it was announced that the Gaia space observatory was ready to begin its operational phase. The commissioning phase had been extended to allow mission controllers to come up with stray light mitigation strategies

Gaia's deployed mass is about 2000 kilograms. It is about 3.5 meters in length and has a diameter of 10 meters. The disk-like structure in Figure 19.2, which surrounds the cylindrical telescope, is the stray light shield.

One requirement for Gaia's successful operation is ultraprecise pointing capability. This is achieved using gyroscopes without moving parts, star trackers, and Sun sensors.

The umbrella-like 10-meter diameter sunshade was folded during launch and deployed early in the mission. One purpose of this unit is to protect the spacecraft's light-sensitive payload unit from direct exposure to sunlight. Another function is to keep the temperature of instruments as constant as possible to reduce the contraction and expansion of metal parts caused by temperature variation. Such variations would slightly affect instrument geometry and degrade image quality if they were not mitigated.

Figure 19.2. Artist's conception of the Gaia spacecraft on station (courtesy ESA).

After the sunshield deployed, six rectangular gallium arsenide solar panels were unfurled. These have a total surface area of 12.8 square meters and always face the Sun. At mission's end, they will still provide 1.91 kilowatts of solar power.

Some of this power is routed to the Command and Data Management Unit. This controls all functions of the Gaia spacecraft, including communication, navigation, onboard power distribution, and payload operation.

Housekeeping and science data are stored in a solid state mass memory system. This device has a capability of 800 gigabits. Data are stored in this unit and prepared before they are transmitted to ground stations by medium and low-gain antennas.

THE GAIA INSTRUMENT SUITE

There are three instrument units aboard this spacecraft. These are the Astrometric Instrument, the Radial Velocity Spectrometer, and the Photometric Instrument.

The purpose of the Astrometric Instrument is to study and determine star position, distance, and motion across the sky (proper motion). The Radial

Velocity Spectrometer uses high-resolution stellar spectra to measure radial motion (motion toward or away from Gaia). The Photometric Instrument is used to create low-resolution stellar spectra in the blue and red spectral ranges; data from this instrument reveal stellar mass, temperature, and chemical composition.

Gaia spins at four revolutions per day. Many stars can be viewed during each spin period by the two onboard telescopes. To produce an all-sky survey, the satellite is put into a 63-day precession cycle. On average, each star that passes through the field of view of an onboard telescope can be mapped 70 times during a precession cycle.

The two telescopes in Gaia's instrument suite share an optical bench and a common focal plane assembly. The primary mirrors of these telescopes are 0.5 and 1.45 meters in diameter. Light received by the two mirrors can be combined so that the resulting images can be superimposed and viewed simultaneously. As is true of many amateur and most professional telescopes, starlight falls on an array of charge-coupled devices (CCDs) at the telescope's focal plane. Sensitivity can be adjusted to allow viewing bright or dim stars.

During the 5-year operational mission phase, data processing on Earth should allow the Astrometric Instrument to measure the angular position, proper motion, and parallax of thousands of stars simultaneously. Parallax shift, which is used to calculate star distance, is the angular shift of a relatively near star against the stellar background caused by Earth's revolution around the Sun. One result of this capability is the probable discovery of multiple stars, extrasolar planets, and faint Solar System objects.

At the same time, the Photometric Instrument will be used to generate stellar spectra in the visible and near-infrared spectral ranges. The resulting data on star spectral energy distributions will be useful in determining star properties including surface temperature, mass, age, and composition.

Using Doppler shift measurements from the Radial Velocity Spectrometer, star velocities along the line of sight (toward or away from Gaia) will be obtained for as many as 150 million stars. Star radial velocities will be determined from observations of three infrared emission lines of calcium. This data will be of use in improving our understanding of the motions and evolution of our galaxy.

GAIA AND PARENAGO

Observations from Gaia will certainly be useful in settling the controversy regarding spiral arm density waves discussed in Chapters 17 and 18. However, of greater significance to the subject of this book, accurate proper and radial motion observations of millions of stars in our galaxy should reveal whether Parenago's discontinuity is a local or galaxy-wide phenomenon. If it is evident in stars at distances far in excess of 260 light years, the density wave hypothesis will become far less likely, and the observational evidence in favor of volitional stars will be strengthened.

Although Gaia is designed and intended to observe objects within the Milky Way galaxy, there is certainly nothing stopping it from being applied to study extragalactic objects. If some observing time could be applied to investigating color differences across the spiral arms of external galaxies, the results would be very interesting.

FURTHER READING

Information on the Gaia spacecraft is available online from the ESA's Gaia factsheet (*http://www.esa.int/Our_Activities/Space_Science/Gaia/Gaia_factsheet*).

Information on Gaia's launch and orbit is available online (*http://en.wikipedia.org/wiki/Gaia_(spacecraft)#Launch_and_orbit*).

Information on Lagrange points is available online (*http://en.wikipedia.org/wiki/Lagrangian_point#Spacecraft_at_Sun.E2.80.93Earth_L2*).

A recent (July 30, 2014) Gaia mission update is available online (*http://www.spaceflight101.com/gaia-mission-updates.html*).

Gaia's instruments are also described online (*http://www.spaceflight101.com/gaia-science-instruments.html*).

Space.com is a good online source of information on the operational status of Gaia and other robotic and occupied space missions. Another good source is the monthly periodical, *Spaceflight*, which is published by the British Interplanetary Society.

CHAPTER 20

Observational/experimental panpsychism

They sought it with thimbles, they sought it with care;
  They pursued it with forks and hope;
They threatened its life with a railway-share
  They charmed it with smiles and soap.

Lewis Carroll, *The Hunting of the Snark*

Today's scientists have tools a bit more sophisticated than those of Lewis Carroll's Snark hunters. With new instruments like the space telescope Gaia, observational astronomers can hopefully begin to move panpsychism—the doctrine that everything is, to some extent, conscious—from deductive philosophy to science. However, an attempt was made to perform this feat several decades ago.

In 1961 a collection of essays was published on the subject of scientific speculation. One of the authors was Arthur C. Clarke. Renowned both for his contributions to visionary science fiction and space technology, Clarke titled his essay "Trouble in Aquila and Other Astronomical Brainstorms."

Clarke begins this short piece with a discussion of his reluctance to publish half-baked scientific ideas because of his failure to protect his 1945 concepts of 24-hour, geosynchronous orbit communication satellites with a patent or copyright. He then goes on to describe a very simple and provocative experiment, one that can be performed by any amateur astronomer.

Using his copy of *Norton's Star Atlas*, Clarke plotted the celestial positions of bright novae between 1899 and 1936. He noted that they tended to cluster in a small region of the sky, near the constellation Aquila.

Clarke speculated about the possible causes in his essay. He refers to one of his classic science fiction novels, *The City and the Stars*, in which an advanced galactic civilization uses the energy reserves of myriad stars to power an intergalactic migration. If you are a fan of *Star Wars*, you might prefer the hypothesis that the Aquila novae are the result of a war between two technologically advanced interstellar species.

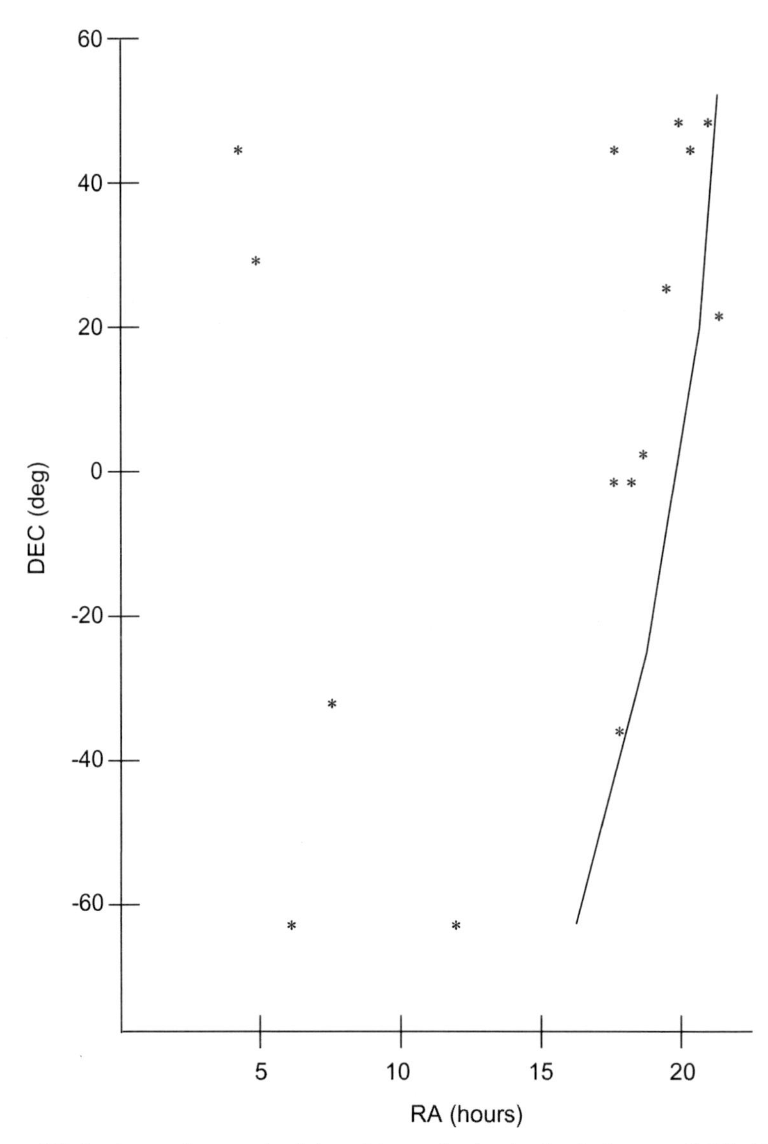

Figure 20.1. Approximate celestial positions of galactic classical novae (1901–1986). The solid line is an approximation of the Milky Way centerline.

However, we are now aware of the fact that Clarke was heavily influenced by the work of Olaf Stapledon. So, it is at least possible that Clarke suspected that he might have stumbled upon the celestial signature of a war between minded stars and unified planetary intellects, as described in Stapledon's novel *Star Maker*.

To check out Clarke's approach, it was necessary first of all to locate the coordinates and names of recent novae. I located them in a table on p. 431 of my copy of *Allen's Astrophysical Quantities*. These were then plotted on a chart of sky position, which is presented as Figure 20.1. In this figure, declination (DEC) is the celestial equivalent of latitude and right ascension (RA) is the celestial equivalent of longitude. Declination is in degrees, right ascension is in hours.

It is interesting to note that the novae positions do indeed cluster, as Clarke reported. To further investigate the significance of this clustering, I plotted the approximate centerline of the Milky Way, as suggested in a website discussing Clarke's essay (*https://www.habitablezone.com/2007/02/19/trouble-in-aquila/*).

When attempting to explain an observational or experimental result, it is wise to apply Occam's Razor and assume that the simplest of suggested explanations is probably the correct one. Yes, it is possible the grouping of novae in Figure 20.1 is the result of an interstellar war. However, there is a simpler explanation.

Most of the novae align closely with the Milky Way centerline. This is a region of high star concentration. So, it is reasonable to assume that the statistics of novae follow the statistics of stars. Where there are more stars, there will be more novae.

However, one set of observations will neither prove nor disprove the volitional star hypothesis. One of these conclusions can be reached only after intense observational study of a number of astrophysical parameters. Since humanity's observational and experimental tools are now up to this task, the balance of this chapter is used to outline some of the phenomena that should be studied. The strategy used is to list a theoretical assumption related to the volitional star hypothesis and then describe how this could be proved or disproved.

ASSUMPTION 1. PARENAGO'S DISCONTINUITY IS A GALAXY-WIDE PHENOMENON

As discussed in Chapter 19, the observational effort to prove or disprove this assumption could be carried out by those charged with analyzing the scientific data on the motions of myriad stars in the Milky Way galaxy received from the new Gaia space observatory.

Demonstration of a general, galactic nature for Parenago's discontinuity will do a great deal to further undermine the spiral arm density waves hypothesis, which is a competitor to the volitional star hypothesis. However, since alternative theoretical concepts will certainly be put forward, such an observational result should be considered supportive of volitional stars but not conclusive.

ASSUMPTION 2. THERE ARE NO SIGNIFICANT COLOR DIFFERENCES ACROSS GALAXY SPIRAL ARMS

Chapter 13 was devoted to a study of stellar motions in the spiral arms of 12 nearby spiral galaxies. To further weaken the case for the spiral arm density waves hypothesis, this already extensive observational program should be extended to include more galaxies. Because of the ~100 billion galaxies in the observable Universe, a sample of a few hundred should be studied to reach a less tentative conclusion.

Spectral data should also be taken across spiral arms, as discussed in Chapter 17. This should be useful in determining whether star color and velocity varies across spiral arms. If there is no variation, the spiral arm density waves hypothesis to explain Parenago's discontinuity is further weakened.

There are many mountaintop and space observatories and others are coming on line in the near future that could contribute to this study. One is the NASA James Webb Space Telescope, which is to be launched in 2018 as a successor to Hubble (Figure 20.2).

Figure 20.2. The James Webb Space Telescope (courtesy NASA).

ASSUMPTION 3. UNIDIRECTIONAL STELLAR JETS PLAY A ROLE IN OUTER-GALAXY STELLAR MOTIONS

There are two types of observational evidence which would support the hypothesis that young volitional stars in the outer reaches of our galaxy use unidirectional stellar jets to alter their orbital velocity around the galactic center.

First, the number of these jets should increase as the distance from the galaxy's center increases. This is because stars out to about the Sun's distance from the galaxy's center tend to follow Keplerian orbits.

Second, the direction of jets should be aligned opposite to a young star's galactic trajectory and jet intensity should increase with increasing distance from the galaxy's center.

ASSUMPTION 4. A WEAK TELEKINETIC FORCE IS ONE OF THE PROPERTIES OF CONSCIOUSNESS

As discussed in previous chapters, a weak telekinetic force intrinsic to consciousness might influence the motions of volitional stars in the outer reaches of a galaxy and at least partially explain the kinematics anomaly. This hypothetical force would be many orders of magnitude weaker than that necessary to perform the magician's trick of utensil bending by telekinetics.

A means by which quantum physicists and parapsychologists could search for this force and quantify it was suggested by a respondent to my Centauri Dreams blog piece. Subjects could remotely observe a Bose–Einstein condensate and will it to climb the walls of its container to a specified height. Their scores would be determined by the number of times (if any) this height was achieved over a specified time.

ASSUMPTION 5. CONSCIOUSNESS IS A FUNCTION OF MOLECULAR LEVEL QUANTUM EVENTS AND COMPLEXITY

Computers are growing in complexity and shrinking in size. If consciousness does arise from quantum events in molecular systems and if it increases with system complexity, our computational devices should at some point begin to demonstrate conscious behavior. I have no idea how this behavior could be measured or even how it would be manifested. Perhaps if we ask or request our conscious computer to perform a task and it is not in the mood, it will respond, "not tonight honey, I have a headache."

CONCLUSIONS

The difference between the way in which universal consciousness is now considered and the past is that modern scientists are in a position to gather supporting (or falsifying) evidence. It is remarkable that so many in the past have speculated about these concepts, before the observational and experimental tools to study them were available. For instance, in the same 1961-vintage volume that contains Arthur C. Clarke's essay on novae, A. D. Maude speculates that sunspots have many of the attributes of living organisms.

FURTHER READING

Arthur C. Clarke's essay is included in L. J. Good (ed.), *The Scientist Speculates: An Anthology of Partly-Baked Ideas* (Basic Books, NY, 1962). My approximate plot of novae positions used data from table 17.1 of W. M. Sparks, S. G. Starrfield, E. M. Sion, S. N. Shore, G. Chanmugam, and R. F. Webbink, "Cataclysmic and symbiotic variables," in *Allen's Astrophysical Quantities*, 4th edn., chap. 17 (ed. A. N. Cox, Springer-Verlag, NY, 2000).

Information on the James Webb Space Telescope is available online (*http://jwst.nasa.gov/index.html*).

CHAPTER 21

Can we talk?

> In the midst of the word he was trying to say,
> In the midst of the laughter and glee,
> He had softly and suddenly vanished away—
> For the Snark *was* a Boojum, you see.
>
> Lewis Carroll, *The Hunting of the Snark*

The basic thrust if this book is the search for kinematic aspects of stellar consciousness. These would require only a very primitive form of self-awareness among the stars, a herding instinct similar to that of the slime mold amoeba.

But, what if the stars are indeed godlike as in the science fiction novel of Gregory Benford and Gordon Eckland discussed in Chapter 11? How do we deal with such entities and avoid the disastrous war between planetary and stellar entities in Olaf Stapledon's *Star Maker*?

If stars are truly minded and humankind fails in its attempts to resolve differences between peoples, planetary intelligence will almost certainly not fade out like Lewis Carroll's unfortunate Snark hunter; if the Sun is a Boojum, we will likely go out with a bang!

As borne out by history, humans have not been very successful in resolving differences with other humans, let alone truly alien conscious life forms. We go to war at the drop of a hat, use innocents as human shields, gleefully commit ethnic cleansing, and naively place young men and women in positions where they will almost certainly become proficient at the high art of torture.

Furthermore, our record with our nearest nonhuman relatives has not been very inspiring. Yes, we now attempt to communicate with and protect many varieties of cetaceans, the order of nonhuman mammals with the highest brain/body mass ratio. But about a century ago, we almost hunted them to extinction.

We have domesticated and formed partnerships with some canines, equines, and felines, but their wild relatives are in many cases threatened by human encroachment. Our record with the highly intelligent and social elephant has been decidedly mixed.

Some species of birds, descendants of the mighty dinosaurs that ruled the Earth for 200 million years, have been reduced to pets; others lead horrible and short lives on factory farms. Even many of our closest relatives, apes and other primates, are hunted for bush meat and threatened with extinction.

So, those who conclude that our ability to communicate and empathize with nonhumans is laughable may have a point. Nevertheless, humans have the capacity to evolve and change. Many people in the developed world are aware of the social and biospheric problems discussed here. They are developing a more tolerant and creative attitude toward other humans and other forms of life. As concepts of nonviolence and pacifism become more widespread and interest in vegetarian and vegan lifestyles spreads, our species may mature in its attitude toward nonhuman consciousness.

However, communicating with such an entity as a minded star will present major challenges. This will be especially true of stellar minds which are as far removed from individual human minds as humans are from the hive minds of social insects.

For the past few decades, a small band of SETI (search for extraterrestrial intelligence) astronomers have been investigating the possibility of communicating with nonhuman, nonterrestrial conscious entities. The experience of SETI scientists may be invaluable for those hoping to communicate with the stars.

BASIC ASSUMPTIONS OF SETI

If humanity is not the youngest of the technological civilizations in the galaxy, it is certainly one of the youngest. Seeking minds among the stars is therefore a very daunting and frustrating experience. Would advanced minds, even minds in organic creatures like us, have a great deal of interest in communicating with us primitives? For example, we spend little time trying to establish a link with sunflowers and gerbils. Yet these organisms share the same basic genetic code with humans, which is most unlikely to be the case with extraterrestrial intelligence.

In the early 1960s the SETI astronomers had difficult choices to make. It was necessary to reduce the number of likely candidate stars from the few hundred billion in the Milky Way galaxy to something more manageable. Then, it was necessary to come up with a probable medium of exchange for communication among widely separated technological civilizations. The next challenge was to devise a suitable language that could be shared by ET and humanity. Finally, a good deal of thought needed to go into the message design. After all, it would not be a good thing if a benign greeting were misinterpreted as a declaration of war!

To reduce the number of stars to a manageable level, it was decided to concentrate on Sunlike main sequence stars (classes F, G, and K on the Hertzsprung–Russell diagram). This is a conservative assumption since we now have discovered habitable planets circling the very numerous M-class red dwarf stars. Many binaries are eliminated from consideration, as are young and old stars. In many binary star systems, habitable planets may have unstable orbits.

Planets of young stars probably have not had time to develop technological life. Planets of old stars may lack the heavy elements necessary for the development of technology. Stars close to the center of the galaxy or likely to pass near a star-forming region are also eliminated from the list of potential SETI candidates because high radiation levels may interfere with the evolution of life on their planets. Finally, most SETI searches concentrate on relatively near stars: those within a few hundred light years. This is because information transfer in the Universe is restricted to light speed; even between relatively near stars a greeting and response would take centuries.

Arriving at the preferred medium of exchange for SETI was a bit easier. Certain possibilities were ruled out immediately. Although telepathy (direct mind-to-mind contact) is a favorite in Olaf Stapledon's *Star Maker*, it has not passed the tests to be considered a reliable information channel, and many scientists do not accept its existence. Although advanced civilizations may use channels such as tachyons (hypothetical faster-than-light particles), neutrinos (massless, nonreacting particles emitted in certain nuclear transformations), or even *Star Trek*'s subspace radio, we have no idea how to utilize such technologies. It was decided that technologically advanced extraterrestrials seeking to contact humans or members of other emerging galactic civilizations would use either space travel or electromagnetic emissions. Moreover, electromagnetic approaches such as radio and lasers would be a lot cheaper to implement than interstellar spacecraft.

Finding a suitable language also required some thought. Early SETI scientists were fully aware the electromagnetic spectrum was vast. What would be the most likely wavelength for advanced extraterrestrials to use in their galactic transmissions? It was assumed that water would be necessary for the existence of all organic life in the Universe, as is the case for life on Earth. Since terrestrial life forms often gather around the waterhole, the preferred SETI wavelengths corresponded to the wavelengths of emissions from the hydrogen atom and hydroxyl (OH) radical, since water molecules in the interstellar medium are usually dissociated into these components. These emissions are in the radio range of the spectrum and, therefore, could only be detected using radio telescopes (Figure 21.1). It was obvious that extraterrestrials would not share a common spoken or written language with humans, but their high technology would require mathematics. So, mathematical approaches based on the binary code of computer science were adopted.

Graphic images have been used to make interpretation by extraterrestrials as easy as possible. One of these graphic images has been beamed to a very distant star cluster. Others have been attached to human-launched space probes that have departed the Solar System. In 1974 the huge radio telescope in Arecibo (Puerto Rico) sent a message toward the M13 globular cluster, at a distance of 25,000 light years from Earth. It was designed so that the pulses were the product of two prime numbers (numbers that are products of only themselves and 1). Say an extraterrestrial radio astronomer picked up a signal from Earth consisting of 143 pulses and was aware of the significance of prime numbers;

Figure 21.1. A radio telescope array (courtesy National Radio Astronomy Observatory, NRAO).

ET might arrange them in one of two ways: 11 rows across and 13 columns down, or 13 rows across and 11 columns down. Only one arrangement would result in an information-rich image.

Messages from Earth have been designed to reveal our physical form, size, and galactic location. Moreover, attempts have been made to demonstrate our benign intent. However, these efforts gave not been without controversy. In the early 1970s, NASA attached message plaques to Pioneer 10 and 11 and launched them on extrasolar trajectories (Figure 21.2).

Public discussions of this plaque approved use of the spacecraft image as a representation of the size of typical humans, the representation of human features as a combination of those of many races, and the representation of our Solar System and Pioneer's trajectory. However, some U.S. citizens objected to the notion of sending pictures of nude females to the stars, so the female genitalia were airbrushed out.

Of greater significance to our discussion are the representations of the Sun's position in the galaxy relative to 14 pulsars and the galaxy's center as well as the human male's upraised hand. To most observers, the plot indicating the pulsar position seems harmless; however, one of the people engaged in discussions about the suitability of the plaque mentioned that a race of intelligent

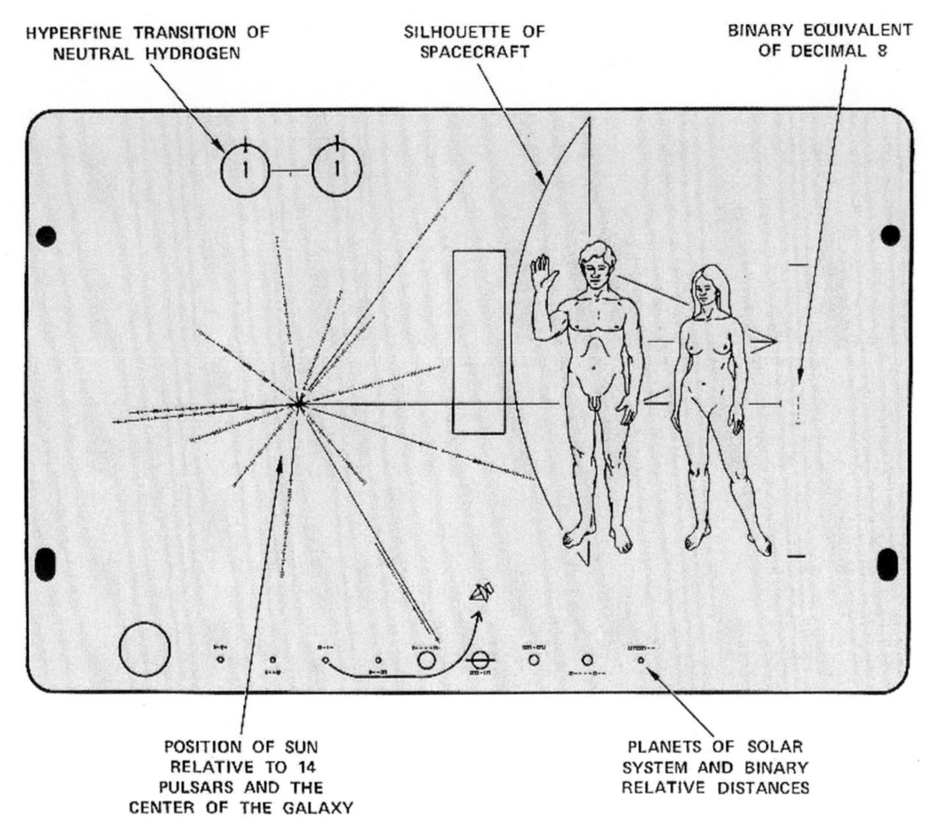

Figure 21.2. The message plaque attached to Pioneer 10/11 (courtesy NASA).

arachnids might interpret the position of this plot as an indication that primates on Earth had enslaved spiders. Another observer pointed out that some species of terrestrial mammals and many reptiles are venomous. Perhaps the male's upraised hand is a warning that he is about to attack using a venom gland in his palm. One species' signal of benign intent might be another species' declaration of war!

STAR TALK

SETI seeks to contact advanced planetary civilizations. These may be organic beings like us, intelligent machines, planetary hive minds, or uploaded virtual beings. However, they will all presumably share the same planetary frame of reference as humanity. At some point in their evolution, they would have developed an understanding of mathematics; otherwise, communication over interstellar distances would not be possible. By attaching message plaques to interstellar spacecraft, the expectation is that extraterrestrials sense their environment in ways similar to terrestrial life forms.

But, when attempting communication with hypothetical conscious stars, all bets are off. Take, for example, the target selection criteria for a communication effort. It might seem reasonable to first attempt communication with our Sun because of its proximity. A message from Earth requires only 8 minutes to reach the solar photosphere. However, in Olaf Stapledon's *Star Maker*, initial communication fails because the stars decide that planetary conscious beings are vermin and should be exterminated. Since we have absolutely no idea of the annoyance tolerance level of a minded star, perhaps it is wiser to attempt communication first with stellar neighbors to the Solar System.

Next, we consider the medium of exchange. Since the Sun (and other stars) are prodigious emitters of electromagnetic radiation, this is likely the best choice. In fact, it might be the only option since no imaginable human space probe would survive very long in close proximity to the photosphere.

Arriving at an appropriate language accessible to both conscious stars and conscious planetary life would be a major challenge. First, there is the difference in life span between a human and a star. A human lives, if lucky, for about 10^2 years. The life expectancy of a Sunlike star is about 10^{10} years. Another way of looking at this is that a human lifetime is the equivalent of about one second in the life of a Sunlike star. I also wonder what possible mutual frame of reference, other than self-awareness, there would be between planetary and stellar consciousness. Perhaps these problems can be dealt with in the future, if the World Wide Web ultimately becomes conscious and develops into a planet-wide hive mind with a lifetime of a billion years.

The same problem arises when considering message content. Would the Sun or beings in the Sun have some analog of mathematics and sight such that they could understand a graphical message?

We may never establish communication with entities as alien as stars. Moreover, this raises some disturbing thoughts. SETI, in its 5-decade history, has listened for transmissions from thousands of stars. In some cases, individual stars have been targeted. In others, all-sky surveys have been used. There have even been unsuccessful searches for the infrared radiation that would be emitted by Dyson/Stapledon spheres constructed by advanced technological civilizations to encircle and capture the light emitted by their stars. With the exception of a few dozen false alarms, the nearby Universe seems to be devoid of intelligent signals since no signal received to date has repeated.

Nevertheless, during the same time frame it has been established that most if not all stars have planets. Many of these planets circle in their stars' habitable zones. In the case of our Solar System, the habitable zone has expanded to include the moons of Jupiter and Saturn. It was formerly supposed that the jump from single-cell life to multicellular life was very difficult. However, the results of recent experiments on stressed yeast cells indicate that this process may not be a limit to the development of advanced life forms. So, is it possible that technological life forms evolve on many habitable planets, begin to rearrange their planetary systems, attract the attention of their host stars, and are exterminated?

FURTHER READING

For a very readable account of the early years of SETI, check out T. R. McDonough, *The Search for Extraterrestrial Intelligence: Listening for Life in the Cosmos* (Wiley, NY, 1987). A more recent treatment of SETI is included in S. J. Dick, *The Biological Universe: The Twentieth Century Extraterrestrial Life Debate and the Limits of Science* (Cambridge University Press, Cambridge, U.K., 1996).

The controversy about the Pioneer 10/11 interstellar message plaque is discussed by Carl Sagan, one of the plaque's designers. See C. Sagan, *The Cosmic Connection: An Extraterrestrial Perspective* (Anchor-Press/Doubleday, Garden City, NY, 1973).

Results of an unsuccessful search for Dyson/Stapledon spheres enclosing nearby stars can be accessed at R. A. Carrigan, Jr., "IRAS-based whole-sky upper limit on Dyson spheres," *Astrophysical Journal*, **698**, 2075–2086 (2009). This paper is available online (*http://home.fnal.gov/~carrigan/infrared_astronomy/Fermilab_search.htm*).

For a summary of the results and implications of the experiment on yeast cells, which apparently organize themselves in multicellular colonies when stressed, see C. Zimmer, "Yeast Experiment Hints at a Faster Evolution from Single Cells," *New York Times* (July 18, 2012). This is available online (*http://www.nytimes.com/2012/01/17/science/yeast-reveals-how-fast-a-cell-can-form-a-body.html*).

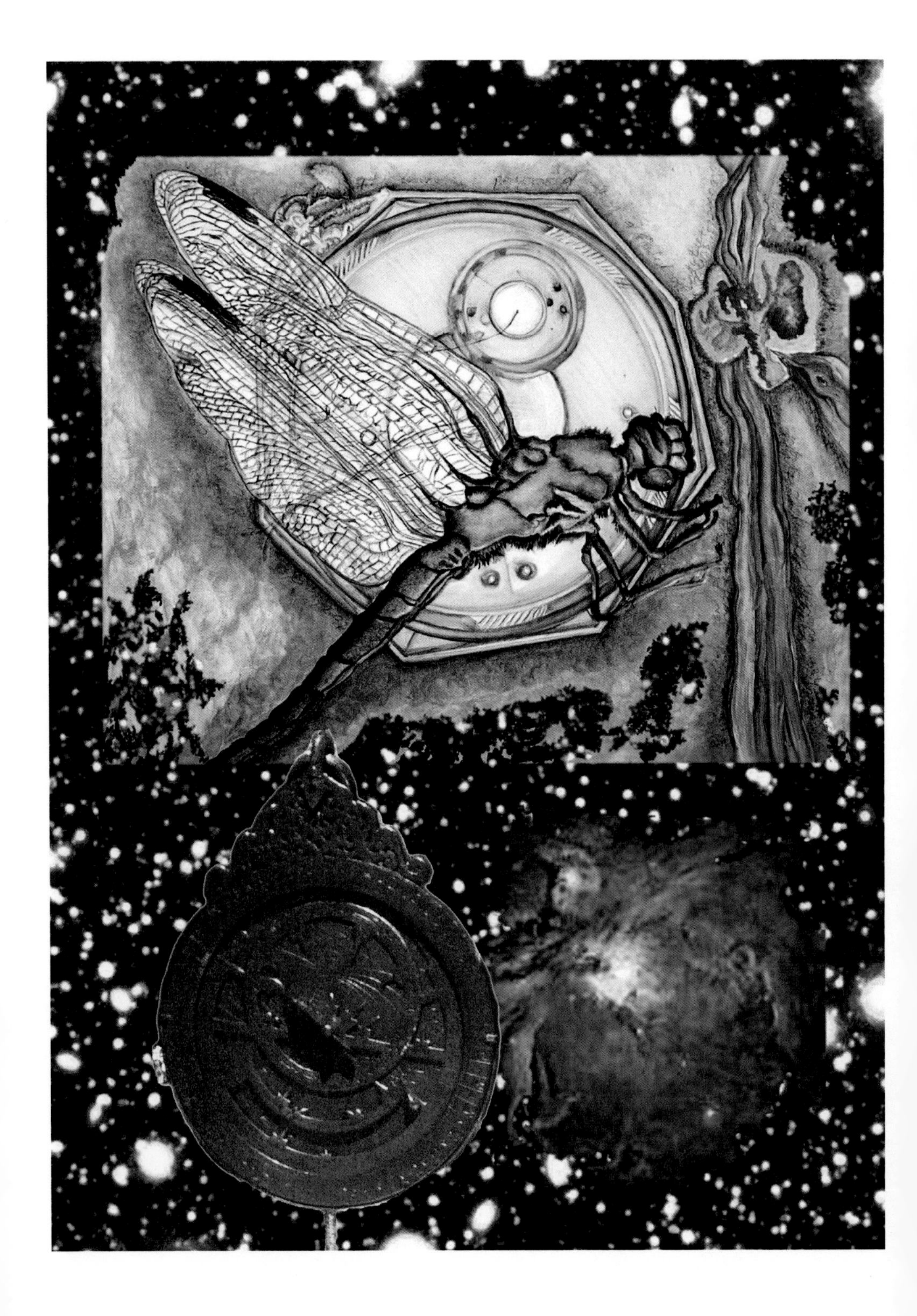

CHAPTER 22

Is the Universe conscious? Might it become conscious?

> The green earth tilts through a sphere of air
> And bathes in a flame of space.
> There are houses hanging above the stars
> And stars hung under a sea
> And a sun far off in a shell of silence
> Dapples my walls for me.
>
> Conrad Aiken, *Morning Song*

What we have been discussing could be the mere tip of a celestial, universal iceberg. Is it possible that the entire Universe, not merely stars and biospheres, is conscious? Is it possible that the Earth, ocean, sky, Sun, and stars are all connected in a universal totality?

Such an interconnectivity of all creation has been the stuff of mysticism for millennia. However, might there be something in such a holistic approach for scientists as well? It is not surprising that the majority of working scientists have adopted a reductionist stance rather than a holistic one. After all, if you are a microbiologist it is hard enough to stay abreast of developments in your own field instead of integrating them with the latest news in cosmology. Moreover, when doing experiments it is best to limit the number of variables.

So, it is not surprising that only a minority of scientists have attempted to examine the Universe at many levels, in the hope and expectation of finding some evidence for a universal mind.

One scientist who has attempted this task is Erich Jantsch (1929–1980), an Austrian astrophysicist who spent much of his career working in the United States. Obviously influenced by the Gaia hypothesis, Jantsch applied the principles of systems theory in his arguments that the Universe is self-organizing at all levels.

In his efforts to develop a paradigm of universal self-organization, Jantsch considers nonequilibrium structures. These appear at all levels of his self-organizing Cosmos and are at least partially open to their environment. In Jantsch's view, equilibrium is equivalent to stagnation and death. It is the constant interchange of matter and energy between a stable but never resting nonequilibrium structure and its environment that allows the structure to both renew and evolve. Jantsch denotes such self-renewal as "autopoiesis" and considers structures with this property as possessing a primitive form of consciousness.

Jantsch defines such structures as "dissipative." They import energy from their environment and export entropy (or disorder). Human cultures and civilizations may be considered as complex types of dissipative structures. As it evolves, an isolated human culture increases its complexity and the amount of information it possesses. However, this is accomplished at the expense of environmental degradation. According to the second law of thermodynamics, even the greenest human civilization will unavoidably degrade its environment to a certain extent.

As the Universe evolves from one state to another, an additional function encourages self-organization. This is symmetry breaking. Very shortly after the Big Bang, time asymmetry became significant with the break between past and future, which is involved with causality and temporal order. Another example of symmetry breaking in the early Universe is the emergence of the four fundamental forces (gravity, electromagnetism, nuclear weak, and nuclear strong) from an original "superforce" as the early Universe cooled in the nanoseconds after the Big Bang.

SELF-ORGANIZATION AT THE MOLECULAR LEVEL

One pre-organic example of self-organization is the formation of crystals. As conditions permit, these solid structures emerge from the liquid state with more form and higher entropy than the liquid from which they develop.

Jantsch also discusses another dissipative structure at the molecular level, the Belousov–Zhabotinsky reaction, which is named for the Russian scientists who studied it. If conditions are right and one oxidizes malonic acid by a sulfuric acid solution containing bromate, in the presence of cerium, iron, or manganese ions, interference patterns formed by rotating spiral or concentric waves are observed. Pulsations of great regularity, some of which occur over periods of many hours, have been observed and recorded. It seems as if dissipative structures of this type renew themselves and attempt to maintain a stable space-time structure.

Dynamic chemical structures of this type are certainly not as conscious as humans and other higher animals. Nevertheless, it seems reasonable to assume that their apparent "interest" in maintaining self-renewal and integrity represents a primitive form of consciousness, perhaps a manifestation of Penrose's universal field of proto-consciousness.

SOME SIMPLE ORGANIC EXAMPLES OF SELF-ORGANIZATION

Jantsch continues his investigation by considering self-organization at a somewhat more complex level of development: that of the single-cell organism. Placed in an aqueous environment with more nutrients (or environmental disturbances) on one side than the other, single-cell organisms such as certain types of algae will orient themselves spontaneously to take advantage of the situation. Algal communities seem to engage in self-organization as they do this.

Another simple organism of interest in self-organization is the slime mold amoeba. These single-cell organisms spend part of their existence as single entities. However, under appropriate conditions they will self-organize into cooperative, sluglike creatures that will crawl through the forest, erect a tower many amoeba high, and eject reproductive cells from that edifice.

At a somewhat higher level of organic development, there is the behavior of hive insects such as ants and bees. Networks of neurons also seem to demonstrate self-organizing behavior.

Beginning at least at the level of simple life, a macroscopic indeterminacy analogous to Heisenberg's microscopic uncertainty principle seems to be built in to the process of evolution. It is almost as if nature has a tendency to produce more novelty and complexity as time progresses.

Evolution also seems to have different branches. Microscopic evolution was most active in the young Universe. Macroscopic evolution becomes more significant as the Universe matures.

Biological systems, even very simple ones, seem to have properties that favor the development of dissipative structures. First, they are linked to the environment by the exchange of energy, which permits maintenance of the biological structure far from equilibrium. Next, there are many chemical and diffusion phenomena that depend upon nonlinear factors operating at the molecular level. Finally, there is nonequilibrium in terms of matter interchange with the environment as well as energy.

COEVOLUTION AS SELF-ORGANIZATION

As Jantsch points out, the concept of coevolution owes a lot to the work undertaken by two American biologists, Paul Ehrlich and Peter Raven, in 1965. There are certain mutual adaptations of disparate species that are very hard to explain by the precepts of orthodox Darwinian evolutionary theory.

One example is the arrangement between certain plant species and caterpillars. Some plants contain significant amounts of alkaloids, which are poisonous to most caterpillars. A few species of caterpillars, such as that of the Monarch butterfly, are immune to alkaloid poisoning so they can safely eat the plant. Many predators soon learn to avoid Monarch butterflies since a meal consisting of one could be uncomfortable or fatal. Returning the favor to the plant, the mature insects contribute to pollination. Jantsch cites studies of other predator–prey relationships while making his case for coevolution. Some of these affect entire ecosystems, as is the case in Australia where fires started by Aboriginal tribes not only cleared land for human use, but increased soil fertility.

Another familiar example is the coevolution of hummingbirds and orchids. The beaks of some hummingbirds are specialized such that they are the primary pollinators of certain species of orchids. Coevolution became a hotly discussed subject resulting in a new journal being published. Between 1974 and 1985, the *CoEvolution Quarterly* was issued, as a spin-off of Stewart Brand's *Whole Earth Catalog*.

SELF-ORGANIZATION AT THE COSMIC LEVEL

Jantsch notes that galaxy clusters emerge first in the young Universe, followed by quasars, which become supermassive black holes at the center of spiral galaxies. Within young galaxies, massive stars evolve quickly and explode as supernovae, seeding interstellar clouds with the heavy atoms necessary for the development of rocky planets and organic life. Inside mature stars similar to the Sun, the thermonuclear reaction rate seems well balanced to insure star stability long enough for the evolution of higher life on planets in the habitable zones of those stars.

In keeping with the central thesis of this book, Jantsch speculates on stellar consciousness. He notes that stars, unlike dissipative structures, obtain energy from thermonuclear processes deep in their interior, not from interchange with the environment. However, the convection layer just below the stellar photosphere, in which energy is percolating upward to eventually be radiated into space, might constitute a dissipative structure. Stars do seem to evolve over long periods of time in a manner that seems to optimize the rate at which matter is transformed into energy. Like dissipative structures at molecular and organic levels, each star demonstrates traits of individuality; it regulates processes and dimensions independent of its environment. One wonders if giant planets such as Jupiter, which radiates more energy than it receives from the Sun, could perhaps have internal layers that might be classified as dissipative structures.

BIOCHEMISTRY AND THE BIOSPHERE

After his consideration of stellar self-organization, Jantsch discusses the role of dissipative structures in pre-biotic evolution and the development of Earth's biosphere. One early result of pre-biotic evolution was the storage of information in nucleic acid molecules, which is a predecessor of the organic genetic code, which stores information in DNA. It is very interesting that pre-biotic and biotic evolution have both operated so as to increase complexity with the progression of time. It is not easy to see how this could occur at random, without a tendency for self-organization built into the system. Information, not matter, is the basic commodity of organic evolution, as opposed to the changes in matter at various stages of cosmic evolution.

Another aspect of biospheric self-organization is the production of Earth's atmospheric oxygen, as a waste product of early life. While rendering Earth's surface unsuitable for these primitive life forms, this atmospheric transformation allowed for the development of many diverse life forms (including humans) which have the ability to thrive in an oxygen-rich environment.

Early on in the evolution of Earth's biosphere, the entire planetary surface seems to organize into a self-regulating system that operates to perpetuate the spread and further evolution of organic life. Consideration of the various feedback loops involved in this self-regulation gave rise to the Gaia hypothesis.

Jantsch gives many examples of novel evolutionary devices that arise in the biosphere with the passage of time: sexuality, the tendency of life to feed on life, the development of multicellular organisms, etc. Evolution might be viewed as a self-transcendent system: one that reaches beyond its existing boundaries in a creative fashion.

He also devotes a considerable portion of his monograph to sociobiology: the connections between organisms and their environment. There are additional examples of self-organization in Jantsch's book as well as his discussion of the evolution of human societies and communities.

THE BIOSPHERE–COSMOS CONNECTION

All would agree that mind plays a role in the biosphere. If mind also exists as a fundamental element in the Cosmos, then there should be an interconnection between the two realms. Jantsch also investigates this concept.

One such interconnection is the circadian rhythm, or biological tide. The hatching of certain species of fruit fly, for example, is synchronized with light. The death of certain one-day blossoms occurs at sunset. Anyone who has endured a long-duration jet flight knows how difficult it is for the human body to readjust to altered rhythms of the day/night cycle.

Although the Sun seems to be the most significant celestial body in the consideration of biological tides, the Moon is also important. Certain species of oysters open precisely when the Moon is highest in the sky, even if they have been transplanted from distant sites by oyster farmers.

Jantsch cites research indicating that plant growth may be correlated with the 11-year sunspot cycle. And then, of course, there is the correlation between lunar phases and the average menstrual cycle of human females.

Figure 22.1. The Centaurus A elliptical galaxy is recovering from a "recent" meal (courtesy NASA).

Some supporting observational evidence for cosmic self-organization

I could continue for many chapters citing examples from Jantsch's monograph that support the hypothesis of a holistic, self-organizing Universe and a universal field of proto-consciousness. Instead, this chapter will be concluded with a discussion of a supporting astronomical anomaly from another source.

It is well known that large galaxies such as our Milky Way, which contain hundreds of billions of stars, engage in a form of galactic cannibalism (Figure 22.1). Observations indicate that, occasionally, a large galaxy will devour a small, dwarf galaxy with only a few billion stars. It is mysterious that large galaxies can maintain their symmetrical shapes after such repeated gorging episodes without some form of self-organization that works on the galactic scale.

CONCLUSIONS

Having reviewed the case for universal proto-consciousness, at least two questions remain. First, if there is no divine Creator or Star Maker, how did a field of proto-consciousness permeate the early Universe? Second, could a fully conscious Universe evolve, as discussed by Olaf Stapledon?

In Stapledon's novels, consciousness is built into the fabric of the Universe as a detached part of the Creator. However, speculative modern physics may allow us to dispense with divinity.

Recall that in Penrose's theory of consciousness, which is discussed and cited in Chapters 6–8, 11, and 14, Bose–Einstein condensate layers in neutron stars may be conscious. As a supermassive star evolves toward the final black hole state, it may first pass through the neutron star stage.

Lee Smolin, an American theoretical physicist associated with the Perimeter Institute for Theoretical Physics, the University of Waterloo, and the University of Toronto, has proposed a cosmological theory based upon natural selection and black holes. He suggests that collapsing black holes result in the emergence of new universes. Universes capable of producing large numbers of black holes are those most likely to reproduce.

If a neutron star becomes conscious as it contracts toward the black hole phase, perhaps this field of conscious information is preserved. It is then altered to become the proto-consciousness field of the daughter universe.

To address the second question, recall our earlier discussion of Stapledon's *Star Maker*. In Stapledon's vision, the Creator of the Multiverse judges the qualities of integrated consciousness that ultimately arise in the various created universes. Unless the speed of thought is demonstrated to be far faster than light, it is difficult to understand how such a universal consciousness could develop. It would take billions of years for a single thought to permeate all portions of a Universe-spanning brain. Perhaps the only way around this would be for future planetary and stellar intelligences across the Universe to band together and produce a Universe-spanning network of wormholes to allow the more rapid spread of information. Whether such a far future prospect is physically or technologically possible is unknown at the present time.

FURTHER READING

Perhaps Erich Jantsch's most influential work is E. Jantsch, *The Self-Organizing Universe: Scientific and Human Implications for the Emerging Paradigm of Evolution* (Pergamon, NY, 1980). This is an expanded version of lectures presented at University of California, Berkeley in May 1979. Various websites discuss Jantsch's contributions. A good one is "The Evolutionary Vision of Erich Jantsch" (*http://erichjantsch.blogspot.com/*).

There are many online sources describing the mutual adaptations of certain species of hummingbirds and the orchids they pollinate. A good one is C. Siegel, "Orchids and Hummingbirds: Sex in the Fast Lane. Part 1 of Orchids and Their Pollinators" (*http://www.fs.fed.us/wildflowers/pollinators/documents/orchids_hummingbirds.pdf*).

The history of *CoEvolution Quarterly* is available online (*http://en.wikipedia.org/wiki/CoEvolution_Quarterly*).

One astrophysicist who has investigated the mysteries of galactic cannibalism and galactic mergers is my colleague Ari Maller. See A. Maller, "Halo Mergers, Galaxy Mergers, and Why Hubble Type Depends on Mass," presented at *For-*

mation and Evolution of Galaxy Disks, Vatican Observatory, Rome, October 1–5, 2007.

The work of Lee Smolin is available online (*http://en.wikipedia.org/wiki/Lee_Smolin*). His theory of cosmological natural selection is summarized in a popular book, *The Life of the Cosmos* (Oxford University Press, Oxford, U.K., 1999).

CHAPTER 23

Expanding paradigms

> Full fathom five thy father lies:
>   Of his bones are coral made;
> Those are pearls that were his eyes;
>   Nothing of him that doth fade
> But doth suffer a sea-change
> Into something rich and strange.
>
> William Shakespeare, *A Sea Dirge*, from *The Tempest*

When anomalies reach a certain level, the requirement for radical change in scientific paradigms becomes unmistakable. In an August 2004 report to the U.S. Air Force Research Laboratory dealing with the physics of teleportation, my friend and colleague Eric Davis discusses the history of previous paradigm modifications in the discipline of physics.

These alterations in mainstream physical thought seem to occur when anomalies build up to the point when they can no longer be ignored. Most mainstream physicists naturally resist such changes as long as they can. Established workers in the field have a vested interest in preserving their funding base. Younger graduate students and postdoctoral researchers do not wish to see years of effort discarded by a radical shift in physical thought.

So, it is to be expected that the inertia in the astrophysics community regarding possible alternatives to dark matter, for example, will not simply disappear. However, there will come a time, if current explanations and funded searches for this phenomenon continue to bear null results, when the direction of astrophysical thought must start to change.

Whether the approach presented in this book bears fruit is, of course, an unknown. It may be that the concept of volitional stars is too radical an approach to be accepted by mainstream physics. However, it may also be that it is not radical enough. To quote Niels Bohr's response to a colleague's attempt to propose a solution to a quantum anomaly during the early 20th century,

> "We are all agreed that your theory is crazy. The question which divides us is whether it is crazy enough to have a chance of being correct."

In physics, new paradigms must incorporate the experimental results obtained in the past. Therefore, I preferred to title this chapter "Expanding paradigms" rather than the more commonly discussed "Shifting paradigms."

My strategy in researching and writing this book has been to work intensively during summer and early autumn of 2014. Then, I turned the draft manuscript over to C Bangs so that she could provide some light editing, suggest refinements, and incorporate her art as chapter frontispieces.

In October 2014, we attended the 65th International Astronautical Congress in Toronto, Canada. During this event, I engaged in a passionate debate with a young colleague regarding the local or nonlocal nature of Parenago's discontinuity. Although nothing conclusive can be said about this issue until observational results from Gaia are received and reduced, I decided to research the matter in greater depth.

Consequently, this final chapter has been added to the text. In addition to reporting further observational data regarding Parenago's discontinuity, it reviews some recent publications on the existence (or nonexistence) of dark matter within our galaxy. The nonastrophysical matters presented here include further discussion of consciousness and the debate regarding the scientific merit of psychokinesis.

I cannot say that the existing physics and astrophysics paradigms will soon be expanded. However, the ongoing debate is a sign of health in the theoretical physics community.

PARENAGO'S DISCONTINUITY AND DISTANT GIANT STARS

As you may recall from preceding chapters, Parenago's discontinuity seems to apply for main sequence stars out to a few hundred light years from the Sun. Although this result supports the conclusion that this effect is nonlocal, the star sample is limited and the maximum star distances are not large enough to rule out local causes.

Within a few years, observations of the motions of millions of stars from the new Gaia European space observatory will hopefully reveal whether Parenago's discontinuity is local or galaxy wide. In the meantime, we can consider the kinematics measurements of giant stars by the Hipparcos space observatory. Because giant stars tend to be more luminous than main sequence stars, some of the giants observed by Hipparcos are very distant.

One researcher who has examined the kinematics of giant stars in the Hipparcos dataset is Richard L. Branham, who is affiliated with the Instituto Argentino de Nivología, Glaciología y Ciencias Ambientales. Some of the giant stars he has considered are thousands of light years from our Solar System. Branham has combined the observational data reduced by him and other workers on giant stars of many spectral classes. In a 2011 paper published in *Revista Mexicana de Astronomía y Astrofísica*, Branham concludes that Parenago's discontinuity exists for giant stars as well as main sequence stars.

In Table 3 of his 2011 paper, Branham lists the motion in the direction of the Sun's galactic rotation (*V*, km/s) relative to the local standard of rest for giant stars of various spectral classes, distance ranges for some of the spectral classes, and sources for these data. To plot a curve of Parenago's discontinuity for giant stars against that for main sequence stars presented in Figure 13.3, it was necessary to estimate (B-V) color indices for the giant star spectral classes in Branham's sample.

First, I accessed a tabulation of coordinates, spectral classes, color indices, and visual apparent magnitudes for 286 bright stars. This table is included on pp. 75–84 of D. A. MacRae, *The Observer's Handbook 1968* (ed. R. J. Northcott, Royal Astronomical Society of Canada, Toronto, 1968). Color indices were averaged for giant stars in this sample corresponding to the spectral class ranges listed by Branham in table 3 of his 2011 paper.

Generally, (B-V) color indices for different stars in a spectral class do not vary much in this sample. An exception was the B1 giant Sigma Scorpii A. The (B-V) color index for this star is listed in MacRae's table as 0.14. Most other stars of this spectral type have slightly negative color indices.

So, it was necessary to dig deeper. I accessed an online version of the Kitt Peak National Observatory Bright Star Catalog, 5th edn. (*http://www-kpno.kpno.noao.edu/Info/Caches/Catalogs/BSC5/catalog5.html*). Bright star HR 6084 matches the coordinates, spectral class, and visual apparent magnitude of this star. Then, I consulted a standard astronomical photometry reference that includes observations from several observatories (H. L. Johnson, R. I. Mitchell, B. Iriarte, and W. Z. Wisniewski, "UBVRI photometry of the bright A stars," *Communications of the Lunar and Planetary Laboratory*, Communication No. 63, Volume 4, Part 3 (University of Arizona Press, Tucson, AZ, 1966). All five of the (B-V) observations of this star, performed at the Catalina Observing Station of the Lunar and Planetary Laboratory cluster around 0.14. Therefore, Sigma Scorpii A remained in the sample.

The reliability of the data in MacRae's tabulation was checked in another manner. A listing of giant star (B-V) color indices is given in table 15.3.1 of J. S. Drilling and A. U. Landolt, "Normal stars," *Allen's Astrophysical Quantities*, 4th edn., chap. 15 (ed. A. N. Cox, Springer-Verlag, NY, 2000). Table 23.1 compares the Drilling and Landolt (B-V) color indices for giant stars of various spectral classes with those calculated by averaging color indices of similar giant stars in MacRae's tabulation. The number of giant stars from MacRae's list used in this comparison for each spectral class is also presented in Table 23.1. Note that the agreement between the two color index estimates is pretty good.

Table 23.2 presents giant star spectral class ranges from Table 3 of Branham's 2011 paper, (B-V) color indices calculated from MacRae's table, giant star motion in the direction of the Sun's galactic rotation (V, km/s) relative to the local standard of rest from Branham's Table 3, the number of MacRae stars used in averaging, distance estimates from Branham or his sources (where available), and the sources of Branham's data. Branham's

Table 23.1. Comparison of Drilling/Landolt and MacRae (B-V) color indices for giant stars of different spectral classes.

Star spectral class	*Drilling/Landolt (B-V)*	*MacRae (B-V)*	*Number of MacRae stars*
G5	0.86	0.86	5
G8	0.94	0.94	4
K0	1.00	1.04	8
K2	1.16	1.20	9
K5	1.50	1.50	3
M0	1.56	1.56	3
M2	1.60	1.63	1

Table 23.2. Giant star spectral class ranges, (B-V) color indices, velocities in direction of galactic center (*V*, km/s), heliocentric distance ranges (1 kpc = 3260 lyr), number of stars in MacRae's sample, and sources for distance estimates.

Giant star spectral class ranges	*Average (B-V)*	*V* (km/s)	*Heliocentric distances* (kpc)	*Star number (in MacRae)*	*Sources for distances*
O-B5	−0.12	8.66 ± 0.38	0.2–3	4	B(2011)
B6-B9	−0.10	11.94 ± 0.41	<0.5	11	B(2009a)
A	0.13	8.00 ± 0.58	<0.5	5	B(2009a)
F	0.34	12.62 ± 0.49	<0.5	2	B(2010)
F-G5	0.69	12.24 ± 0.47	—	6	—
K	1.20	19.11 ± 0.26	<0.3	30	B(2009b)
M	1.59	20.40 ± 0.40	<0.7	9	B(2008)

Distance source key:

B(2008): R. L. Branham, Jr., "Kinematics and velocity ellipsoid of M giants," *Revista Mexicana de Astronomía y Astrofísica*, **44**, 29–43 (2008).

B(2009a): R. L. Branham, Jr., "A study of B6–9 and A giants," *Monthly Notices of the Royal Astronomical Society*, **296**, 1473–1486 (2009).

B(2009b): R. L. Branham, Jr., "Kinematics and velocity ellipsoid of the K giants," *Revista Mexicana de Astronomía y Astrofísica*, **45**, 215–229 (2008).

B(2010): R. L. Branham, Jr., "Kinematics and velocity ellipsoid of the F giants," *Monthly Notices of the Royal Astronomical Society*, **409**, 1269–1280 (2010).

B(2011): R. L. Branham, Jr., "The kinematics and velocity ellipsoid of the GIII stars," *Revista Mexicana de Astronomía y Astrofísica*, **47**, 197–209 (2011).

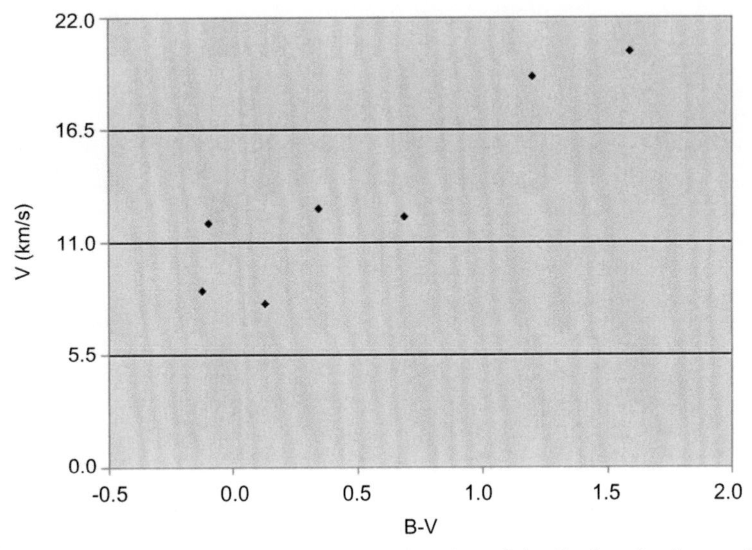

Figure 23.1. Giant star (centroid) motion in direction of the Sun's galactic rotation (V).

sources can be accessed online by searching for the principal author in Google Scholar and downloading the PDF file.

Note from Table 23.2 that high (B-V) giant stars, the cooler stars in the sample, clearly move faster around the galactic center than low (B-V), hotter giant stars. This is in substantial agreement with the results presented for main sequence stars as Figure 13.3.

It is also significant that the heliocentric distances for these giant stars vary from about 600 to 10,000 light years. The typical giant star in Branham's samples is much more distant from our Solar System than the typical main sequence star used in the analysis presented in Figure 13.3. A plot of V versus (B-V) from Table 23.2 for giant stars is presented in Figure 23.1, for comparison with Figure 13.3, which presents a similar plot for main sequence stars.

Even before the new Gaia satellite becomes operational, it seems reasonable to conclude that Branham is correct in his conclusion that Parenago's discontinuity is present in giant stars as well as main sequence stars. Although the results of one study of thousands of stars cannot be considered conclusive, the large distance to typical giant stars in the Hipparcos dataset points toward a nonlocal origin for Parenago's discontinuity. It is too early, of course, to conclude that volitional star motion is the correct explanation, but these results certainly support this hypothesis.

The significance of this result to the hypothesis of stellar volition is that the discontinuity in Figures 13.3 and 23.1 occurs at about the same place in the stellar distribution where molecules become detectable in stellar spectra. This supports the suggestion that consciousness enters matter from a universal proto-consciousness field through the interaction of molecular bonds with vacuum fluctuations. The large distances to many of the stars in the giant star sample argues against local explanations for Parenago's discontinuity.

It should be noted that Table 19.11.1 in *Allen's Astrophysical Quantities*, 4th edn. can also be used to construct a graph which, similar to Figure 23.1, clearly presents Parenago's discontinuity for giant stars. Unfortunately, the distances of the stars in this sample are not listed in that table.

DARK MATTER REVISITED

The hunt for elusive dark matter becomes stranger and more fascinating at every turn. I spent a few weeks in November 2014 surveying some of the latest results in this field.

If a nonlocal cause for Parenago's discontinuity is an anomaly that might lead to an astrophysics paradigm extension, missing dark matter is a much larger anomaly. Many theoreticians, observers, and experimentalists are proposing possible solutions to this enigma. Not all of the proposed solutions are consistent with each other or reliable observational evidence.

For example, an October 16 article in the online digest *ScienceDaily* (*http://www.sciencedaily.com/releases/2014/10/141016085410.htm*) discusses results published by the late George Fraser of Leicester University and colleagues in the October 20, 2014 issue of *Monthly Notices of the Royal Astronomical Society* discussing data received from a European X-ray space observatory. One interpretation of the interesting signal discussed in this paper is that axions, hypothetical but not yet observed subatomic particles, might be produced in the Sun and other stars and ejected into the Cosmos. If these hypothetical cosmic axions exist in sufficient density, they might be the source of the missing cosmic mass.

Not all astronomers accept the axion hypothesis. As discussed in Chapter 12, dark matter is not necessary to explain the galactic motions of the Sun and stars closer to the galactic center. If axions (or some other dark matter candidates) are produced in copious amounts in stellar interiors and ejected into space, one would naturally expect anomalous stellar motions to be most apparent in regions of high star density. However, exactly the opposite effect is observed. As Matt Williams notes in "Is Dark Matter Coming from the Sun?", published online in *Universe Today* (*http://www.universetoday.com/115551/is-dark-matter-coming-from-the-sun/*), some of Fraser's colleagues at Leicester University do not accept the axion interpretation of the strange signal, although they admire the efforts of Fraser's team.

Contradictory results were also reported in "Dark Matter Half What We Thought, Say Scientists," an October 9, 2014 article in *Phys.org* (*http://phys.org/news/2014-10-dark-thought-scientists.html*). According to this report, Prajwal Kafle of the University of Western Australia reached this surprising conclusion by observing star motions near the fringe of our galaxy.

Other researchers, while still accepting the concept that stellar motion anomalies are caused by missing mass, question the assumption that weakly interacting, massive, subatomic particles (WIMPS) are the cause. Adrian Cho in "Dark Matter: Out with the WIMPS, in with the SIMPS?", published on October 30, 2014 in the online version of *Science AAAS* (*http://www.ps.uci.edu/~jlf/research/press/dm_1410science.pdf*), discusses the research of Yonit Hochberg, a theoretician affiliated to Lawrence Berkeley National Laboratory and the University of California. Hochberg and colleagues argue that strange particles that strongly interact with each other, but not normal matter (SIMPS), might form invisible clumps that could produce the anomalies in stellar and galactic motions.

Perhaps the strangest suggestion located by this literature search is described by Charles Q. Choi in "Dark Matter Murder Mystery: Is Weird Substance Destroying Neutron Stars?", a November 18, 2014 article in *Space.Com* (*http://www.space.com/27794-dark-matter-destroying-neutron-stars.html*). According to Choi, Joseph Bramante of the University of Notre Dame and colleagues attempt to explain an unexpected lack of neutron stars near the galactic center by dark matter causing neutron stars to collapse into black holes. Again, is this conclusion consistent with the observational fact that dark matter is apparently not necessary to explain stellar motions closer to the galactic center than that of the Sun?

This short review surveys dark matter suggestions reviewed by online science writers over a period of a few weeks. There are many, many other theoretical possibilities that have been suggested in the more distant past. Even though many of them may seem highly improbable, are any of them "crazy enough" (to paraphrase Bohr's famous quote presented earlier in this chapter) to be correct?

PSYCHOKINESIS DEBATED

If the waters surrounding dark matter are a bit cloudy, those around psychokinesis are positively turbid. On one side, there is substantial literature describing (sometimes) repeatable experiments indicating that some people are able to move small objects from one location to another by the power of will. On the other, there is a well-documented public contest that would award a substantial sum to anyone who could demonstrate such a psychic talent, and thus far the award remains unclaimed. Is there a middle ground in this debate?

In his report to the U.S. Air Force on teleportation physics, Eric Davis concentrates on such aspects of contemporary physics as relativistic wormholes and quantum teleportation of microscopic objects. However, he also devotes a few pages to reviewing the scientific literature on psychic teleportation or psychokinesis.

Davis points out that, during the Cold War, the intelligence agencies and military establishments of countries other than the U.S. evaluated the utility of applying parapsychology. Beginning in the late 1970s and early 1990s, researchers in China conducted and replicated experiments on telekinesis using young subjects. In blind and double-blind experiments, observers reportedly watched as these children and young adults teleported small objects (such as microtransmitters and insects) a distance of meters.

One might be tempted to conclude that these very well-documented Chinese studies demonstrate conclusively that psychokinesis should be treated as a real phenomenon by consciousness researchers. However, there is another side to this debate.

The Amazing Randi, who demonstrated Uri Geller's spoon-bending feat as a magic trick, has been conducting experiments for decades in which alleged psychics are invited to demonstrate their talent. The requirement for winning the One Million Dollar Paranormal Challenge is to demonstrate their talents repeatedly and reliably in eight out of nine trials. After almost 60 years of such public demonstrations, Randi has yet to deliver as much as a penny to an alleged psychic.

How is one to interpret these differing experimental results? It is hard to argue with the rigor of the Chinese experimenters. It is also hard to discount the fact that, over a number of decades, no psychic has collected Randi's million-dollar award.

One research team that has attempted to straddle these extreme interpretations is the International Consciousness Research Laboratories (ICRL), located in Princeton (New Jersey). Formerly known as PEAR (Princeton Engineering Anomalies Research), ICRL is directed by Robert Jahn, a dean and professor emeritus from Princeton University's Engineering and Applied Science division. Brenda Donne is a psychologist who manages ICRL experimental research. The impetus behind this organization's research is the search for small psychokinetic effects that may show up in the function of such systems as high-performance fighter jets. Since these craft are operated by young, highly stressed humans and are equipped with nanoscale electronic systems, such research is not unreasonable. Although one cannot say that ICRL's results to date are conclusive, they may reveal the presence of small-scale psychokinetic effects produced by conscious functioning, not dissimilar from those discussed in Chapter 15 as a possible mode of locomotion for volitional stars.

RECENT PROGRESS IN QUANTUM CONSCIOUSNESS

In November 2014, we attended a lecture at New York University presented by Stuart Hameroff, who is a professor in the Departments of Anesthesiology and Psychology at the University of Arizona in Tucson. Hameroff discussed the quantum consciousness theory that he codeveloped with Roger Penrose of Oxford. I present a summary of some aspects of this theory in Chapter 6.

There have been a lot of recent developments related to this theory. First, a number of research teams have successfully performed experimental tests validating experimental results, discussed in Chapter 6, indicating that quantum effects can occur in small biological structures at ambient temperature. These validations have led various theoreticians to further investigate the core assumptions of the Penrose/Hameroff theory.

Hameroff also discussed a possible clinical application of the theory now under development at the University of Arizona. In this approach, ultrasound is applied to a human subject's frontal cortex. According to the results of the pilot study, the technique can improve subjective mood and assist in the treatment of chronic pain. The authors suggest that application of such transcranial ultrasound acts on intraneuronal microtubules. These may then resonate in the megahertz range. If further studies validate this technique, it may become the first clinical treatment to emerge from quantum consciousness theoretical studies.

LITERARY EXPANSIONS—OLD AND NEW

In late 2014, I learned about additional literary efforts dealing with stellar and universal volition. Surprisingly, one is included in one of the greatest works of fiction. The other is included in a contemporary masterwork of hard science fiction.

While rereading a translation of Leo Tolstoy's 19th century classic *War and Peace*, I encountered a haunting vignette. A group of Imperial Russian soldiers take a break one winter night in 1812/1813 from their pursuit of Napoleon and what remained of the French army. As they sit drinking and eating around a fire, one looks up at the star-filled night sky and remarks that its clarity indicates that there will be frost. When the gathered men return their attention to terrestrial matters and no longer watch the heavens, the stars begin to tremble, vanish, and flare as if whispering to each other. Reading this short passage provokes a similar emotional response to viewing Vincent van Gogh's magnificent *Starry Night*.

Two masters of science fiction, author Larry Niven and author/physicist Gregory Benford, have collaborated to create a two-volume yarn entitled *Bowl of Heaven* and *Shipstar*. In these novels, a ship from Earth encounters a marvelous construct cruising the galaxy: a mobile star with an inhabited, partial shell positioned in its habitable zone. Most of the work considers the interactions among the humans and shell inhabitants. Some characters are beings that live within stars and can direct stellar motions. Other entities include former inhabitants of comets who can communicate with organic and stellar intelligent forms. These novels can be treated as a fictional advertisement for panpsychism.

UNIVERSE AS BRAIN?

In late November 2012, an article by Michael Rundle in the online *Huffington Post* discussed a computer simulation developed by a team of researchers including Dmitri Krioukov of the University of California, San Diego. Their paper in Nature's *Scientific Reports* argued the development and growth of all networks, regardless of size, are governed by the same elusive natural principles. According to this research, the early Universe organized itself in space-time in an analogous manner to the way neurons self-organize within the human brain.

Other complex networks may self-organize in an analogous manner. These include social and biological networks and the Internet. Deepak Chopra in collaboration with Murali Doraiswamy of Duke University, Rudolph E. Tanzi of Harvard University and Massachusetts General Hospital, and Menas Kafatos of Chapman University argue, as a result of the apparent similarity between cosmic and neural networks, that the Universe is actually a brain and organic brains reflect this larger reality.

THE HOLOGRAPHIC UNIVERSE REVISITED

As I mentioned earlier, I am an applied physicist—not a theoretician. However, I am fortunate to teach alongside many very talented theoreticians. One of them, Justin Vazquez-Poritz, a professor of physics at New York City College of Technology, has become interested in the concept of the Universe as hologram. I asked him to summarize his thoughts on the subject and supply a few references. What follows is my digest of his remarks.

First, we need some understanding of the way in which theoretical physicists quantify information. Vazquez-Poritz's model for this involves a large collection of magazines. Each way of ordering this collection—by title, volume, date, subject, etc.—is a microstate of the system defined by the magazine collection. The entropy of the system essentially is a count of all the microstates.

Now, imagine that you put your magazine collection in a room with a volume just large enough to contain it. In this case, the volume of the room is proportional to the entropy of the system.

However, you now come into a bit of money, so you can afford to renovate the room big time. You enlarge it and add floor-to-ceiling bookcases on all four walls. The number of shelves is now just large enough to contain the entire magazine collection. The entropy of the system is now proportional to the area of the four walls in your renovated room—not the volume.

In the 1970s, Jacob Bekenstein began to extend this concept from the domain of architecture to the celestial realm by demonstrating that the entropy of a black hole is proportional to the area of the event horizon—not its volume. This is in agreement with the prediction from general relativity that time freezes at the event horizon from the viewpoint of a distant observer. This result implies that the microstates of a black hole are analogous to an optical hologram, in which light reflected from a two-dimensional surface produces an illusory three-dimensional object. According to the holographic principle, the degrees of freedom of gravitational systems such as black holes can be thought of as existing in the two-dimensional system defined by the event horizon.

This dimensional "duality" is analogous to the fact that the wave and particle aspects of quantum systems can be observed independently depending upon the measuring apparatus. In more conventional terms, whether a three-dimensional or two-dimensional "reality" is observed is like the familiar story of the three blind men describing an elephant, or what would happen if a sighted and a blind art critic were asked to describe the same painting. All observers might be correct, but each might see only a fraction of the object under scrutiny. The blind art critic is sensitive to the two-dimensional aspects of the painting such as surface roughness and texture, whereas the sighted art critic sees a three-dimensional image.

This lends credence to the concept of a self-organizing Universe if, like a monochromatic optical hologram, each tiny part of the holographic Universe contains all of the information of the whole. But, what does it imply for the notion of volitional stars?

In the 1990s, Gerard 't Hooft and Leonard Susskind proposed the holographic principle in the context of gravity. Juan Maldacena then discovered that the duality could be specifically defined for certain gravitational backgrounds using multi-dimensional string theory. Maldacena's approach, which is known as anti-de Sitter/conformal field theory (AdS/CFT) correspondence, has not yet been verified, but it implies that a black hole observed in a space-time of five dimensions offers a holographic representation of a star in four dimensions. As noted in Chapter 6, the Penrose/Hameroff quantum consciousness theory has been supported by replicable experiments and yielded a possible clinical application, which implies neutron stars are conscious. Since neutron stars are a necessary step in the evolution of a massive star toward the black hole state, information about a conscious neutron star might be spread around the event horizon of the black hole—not lost in its interior. If every star has a corresponding conscious black hole in a higher dimension, this interpretation of string theory lends support to the concept of volitional stars.

CONCLUSIONS: SUMMING THINGS UP

We have covered a lot of ground in this chapter. It has been demonstrated that, according to one observational survey, Parenago's discontinuity appears to be nonlocal. This matter cannot be considered resolved because of the limited number of stars in the sample. Hopefully, the newly launched Gaia satellite will add greatly to our knowledge of stellar kinematics.

The search for dark matter as an explanation for anomalies in stellar and galactic motions remains inconclusive. Some would suggest that it is indeed muddy!

Not all scientists reject the concept of psychokinesis. Studies funded by the U.S. Air Force (available online) reveal the existence of allegedly replicable Chinese experiments indicating the existence of this phenomenon. Nevertheless, the matter remains unsettled and hotly debated.

The most highly developed theory of quantum consciousness has received some experimental validation. A possible clinical application of this theory is under development and has been described in the peer-reviewed literature.

The concept of stellar volition continues to crop up in the literature. This includes classic works of fiction and science fiction novels coauthored by respected physicists.

Surprisingly, not only does quantum physics leave open the possibility of conscious stars, but string theory, one of the hottest subfields of theoretical physics, is open to this possibility and to the possibility of a self-organizing, holographic Universe.

However, it is very premature to claim that this book conclusively proves the validity of either volitional stars or a self-organizing Universe. Nevertheless, if the book helps elevate the debate between those who believe in epiphenomenalism and those who believe in panpsychism from the realm of deductive philosophy to the realm of observational and experimental science it has succeeded.

FURTHER READING

For a review of pre-2004 thinking about the physics behind the teleportation of physical objects as a result of applying theoretical and experimental results of quantum theory and relativity, see E. W. Davis, "Teleportation Physics" (AFRL-PR-ED-TR-2003-0034, special report to the U.S. Air Force Research Laboratory, Edwards Air Force Base, CA, 2003; *http://fas.org/sgp/eprint/teleport.pdf*). In this report, Davis considers the classical and quantum theoretical basis for the remote movement of microscopic and macroscopic objects through space-time. Quantum teleportation theory and experimental results and traversable wormholes, which are possible according to relativity theory, are discussed as well as psychokinesis.

The famous quote of the quantum pioneer Niels Bohr, given in the text, is available online (*http://www.goodreads.com/author/quotes/821936.Niels_Bohr*).

Richard L. Branham's paper on giant star kinematics and Parenago's discontinuity for these stars is R. L. Branham, "The kinematics and velocity ellipsoid of the GIII stars," *Revista Mexicana de Astronomía y Astrofísica*, **47**, 197–209 (2011), which is also available online (*http://www.astroscu.unam.mx/rmaa/RMxAA..47-2/PDF/RMxAA..47-2_rbranham.pdf*).

A recent review of The Amazing Randi's One Million Dollar Paranormal Challenge is A. Higginbotham, "The Unbelievable Skepticism of the Amazing Randi," in the November 9, 2014 online edition of *The New York Times Maga-*

zine (*http://www.nytimes.com/2014/11/09/magazine/the-unbelievable-skepticism-of-the-amazing-randi.html?_r=0*).

Information about the experimental and theoretical research performed by ICRL and PEAR, including a list of technical and popular publications by members of the research team at the organization, is available online (*http://icrl.org/*).

One team that experimentally verified quantum effects in one-celled organisms at ambient temperature after studying photosynthesis in algae is E. Collini, C. Y. Wong, K. E. Wilk, P. M. G. Curmi, P. Brumer, and G. D. Scholes, "Coherently wired light-harvesting on photosynthetic marine algae at room temperature," *Nature*, **463**, 644–647 (2010). The abstract of this paper is available online (*http://www.nature.com/nature/journal/v463/n7281/full/nature08811.html*).

Other fairly recent experimental and theoretical work relating to the Penrose/Hameroff theory is described by N. E. Mavromotos, "Quantum coherence in (brain) microtubules and efficient energy and information transfer," *Journal of Physics: Conference Series*, **329** (article 01206). This paper was presented at the *Ninth International Frohlich Symposium: Electrodynamic Activity of Living Cells (including Microtubule Coherent Modes and Cancer Cell Physics)* and is available online (*http://iopscience.iop.org/1742-6596/329/1/012026*).

For more information on the transcranial ultrasound technique, see S. Hameroff, M. Trakas, C. Duffield, E. Annabi, M. Bagambhrini Gerace, P. Boyle, A. Lucas, Q. Amos, A. Buadu, and J. J. Badel, "Transcranial ultrasound (TUS) effects on mental states: A pilot study," *Brain Stimulation*, **6**, 409–415 (2013).

My edition of *War and Peace* was inherited from my parents. It was translated from the Russian by L. Maude and A. Maude and published by Simon & Schuster in 1942, perhaps in recognition of the heroic deeds of the Red Army who, at great sacrifice, were in the process of destroying the Nazi invaders foolishly treading in the footsteps of Napoleon Bonaparte. Our discussion can be found at the conclusion of chap. 3 of book 15, on pp. 1212–1218 of my edition.

The two-volume science fiction book in which a ship from Earth encounters a marvelous construct cruising the galaxy is G. Benford and L. Niven, *Bowl of Heaven* and *Shipstar* (Tor Doherty Associates, NY, 2012, 2014). The acknowledgements cite both Olaf Stapledon and Freeman Dyson.

An introduction to the concept of similarities in cosmological and neural organization is M. Rundle, "Physicists Find Evidence that the Universe is a 'Giant Brain'" (*http://www.huffingtonpost.co.uk/2012/11/27/physicists-universe-giant-brain_n_2196346.html*).

The scientific paper describing the computer simulation leading to the realization of this similarity is D. Krioukov, M. Kitsak, R. S. Sinkovits, D. Rideout, D. Meyer, and M. Boguna, "Network cosmology," *Nature Scientific Reports*, **2**, article 793 (November 16, 2012), which is available online (*http://www.nature.com/srep/2012/121113/srep00793/full/srep00793.html*).

For the implications of this network similarity to panpsychism, see D. Chopra, M. Doraiswamy, R. E. Tanzi, and M. Kafatos, "Your Brain is the Universe—Part 1," *Huffington Post* (June 1, 2013), which is available online (*https://www.deepakchopra.com/blog/view/1120/your_brain_is_the_universe_part_1*).

Justin Vazquez-Poritz was kind enough to supply some technical references on the holographic principle:

J. D. Bekenstein, "Black holes and entropy," *Physical Review D*, **7**(8), 2333–2346 (1973).

S. Hawking, "Black hole explosions?" *Nature*, **248**(5443), 30–31 (1974).

C. R. Stephens, G. 't Hooft, and B. F. Whiting, "Black hole evaporation without information loss," *Classical and Quantum Gravity*, **11**(3), 621 (1994).

L. Susskind, "The world as a hologram," *Journal of Mathematical Physics*, **36**(11), 6377–6396 (1996).

J. M. Maldacena, "The large N limit of superconformal field theories and supergravity," *Advances in Theoretical and Mathematical Physics*, **2**, 231–252 (1998).

O. Aharony, S. S. Gubser, J. M. Maldacena, H. Ooguri, and Y. Oz, "Large N field theories, string theory and gravity," *Physics Reports*, **323**, 183 (2000).

G. Policastro, D. T. Son, and A. O. Starinets, "The shear viscosity of strongly coupled N = 4 supersymmetric Yang–Mills plasma," *Physical Review Letters*, **87**, 081601 (2001).

C. P. Herzog, A. Karch, P. Kovtun, C. Kozcaz, and L. G. Yaffe, "Energy loss of a heavy quark moving through $N = 4$ supersymmetric Yang–Mills plasma," *Journal of High Energy Physics*, **0607**, 013 (2006).

CHAPTER 24

Appendices

APPENDIX 1: PREFIXES

Prefixes are combined with units to denote multiplication. For instance, the term "kilometer" means "1000 meters" or "10^3 meters" and is abbreviated "km". The term "millimeter" means "0.001 meters" or "10^{-3} meters" and is abbreviated "mm". Here are some of the more common prefixes:

Prefix	*Multiplication factor*	*Abbreviation*
giga	$10^9 = 1{,}000{,}000{,}000$	G
mega	$10^6 = 1{,}000{,}000$	M
kilo	$10^3 = 1{,}000$	k
centi	$10^{-2} = 1/100$	c
milli	$10^{-3} = 1/1{,}000$	m
micro	$10^{-6} = 1/1{,}000{,}000$	μ
nano	$10^{-9} = 1/1{,}000{,}000{,}000$	n

Source: H. C. Ohanian, *Physics*, 2nd edn. (Norton, NY, 1989).

APPENDIX 2: SCIENTIFIC NOTATION

Scientific notation, also called "exponential notation" and "powers of 10," is a kind of mathematical shorthand. Its purpose is to convert large and small numbers into approximate "counting numbers" (between 1 and 10) and eliminate errors that would be made in writing lots of zeroes. It also works to convert multiplication between large and small numbers into addition and division into subtraction, also with the purpose of reducing the possibility of errors.

Consider, for example, a large number in conventional notation: 2,700,000. In scientific notation, this is written 2.7×10^6, where "2.7" is the approximate counting number and "6" is the exponent or power of 10. The power or exponent is the number of places the decimal point must be moved to convert the number to an approximate counting number. For numbers smaller than 1, the exponents are negative. Here are some examples:

$$55{,}000{,}000{,}000 = 5.5 \times 10^{10}$$

$$100{,}000{,}000 = 1 \times 10^{8}$$

$$557{,}000 = 5.57 \times 10^{5}$$

$$300 = 3 \times 10^{2}$$

$$7 = 7 \times 10^{0}$$

$$0.05 = 5 \times 10^{-2}$$

$$0.00034 = 3.4 \times 10^{-4}$$

Multiplication: Consider the product: $(4 \times 10^{17}) \times (2 \times 10^{5})$. The product of these two numbers is written: 8×10^{22}.
Now consider the product: $(4 \times 10^{17}) \times (2 \times 10^{-5})$. The product of these two numbers is written: 8×10^{12}.
The rule in multiplying two numbers expressed in scientific notation is to multiply the approximate counting numbers and add the exponents.

Division: Consider the quotient: $(4 \times 10^{17})/(2 \times 10^{5})$. The quotient of these two numbers is written: 2×10^{12}.
Now consider the quotient $(4 \times 10^{17})/(2 \times 10^{-5})$. The quotient of these two numbers is written: 2×10^{22}.
When you divide two numbers expressed in scientific notation, the rule is to divide the denominator's approximate counting number into the approximate counting number of the numerator while you change the sign of the denominator's exponent and then add that to the exponent of the numerator.

APPENDIX 3: UNITS AND CONVERSIONS

Since this book deals with a speculative scientific subject and will (hopefully) be read by residents of Europe, the U.K., and the U.S., units and conversions should be mentioned. In the developed world, most people use the rational metric system. However, the prevailing system in the U.S. is the British imperial system, which dates from the era of divine right rulers. It is an historical accident that the French Revolution, which began the process of sweeping away the divine right rulers, occurred a few decades after the American Revolution.

The following equation presents some metric and British imperial quantities and their conversions. The squiggly equal sign ($\approx$) means "approximately equal". For time units, both systems use "seconds":

$$1 \text{ pound of weight} = 454 \text{ grams of mass at Earth's surface}$$

$$1 \text{ kilogram (kg) of mass} \approx 2.2 \text{ pounds of weight at Earth's surface}$$

$$1 \text{ kilometer} = 1000 \text{ meters (m)} \approx 0.6 \text{ miles of length}$$

$$1 \text{ mile} \approx 1.6 \text{ km}$$

$$1 \text{ meter} = 100 \text{ centimeters (cm)} = 10^3 \text{ millimeters (mm)}$$

$$= 10^6 \text{ microns } (\mu) = 10^9 \text{ nanometers (nm)}$$

$$1 \text{ centimeter} = 2.54 \text{ inches (in.)}$$

$$1 \text{ inch (in.)} = 2.54 \text{ cm}$$

Source: H. C. Ohanian, *Physics*, 2nd edn. (Norton, NY, 1989).

APPENDIX 4: SOME PHYSICAL CONSTANTS

This consideration of some of the constants of nature is arbitrarily divided into two parts: physics and astronomy. All units are in the more commonly used of the two measurement systems: the MKS (meter-kilogram-second) system. This is also called the SI, or standard system of units. The squiggly equals sign means "approximate".

Physics

G = Gravitational constant:	6.67×10^{-11}	Newton-meter^2/kilogram^2
C = Speed of light in vacuum:	3×10^{8}	meters/second
σ = Stefan-Boltzmann constant:	5.67×10^{-8}	watts/m^2-Kelvin^4
h = Planck's constant:	6.63×10^{-34}	Joule-second
m_e = Rest mass of electron:	9.11×10^{-31}	kilograms
m_p = Rest mass of proton:	1.67×10^{-27}	kilograms

Astronomy

1 AU = 1 astronomical unit:	1.5×10^{11}	meters
1 lyr = 1 light year:	$\approx 10^{16}$ meters	$\approx$ 63,000 AU
M_E = Earth mass:	6×10^{24}	kilograms
R_E = Earth radius:	6.378×10^{6}	meters
M_S = Sun mass:	1.99×10^{30}	kilograms
R_S = Sun radius:	6.96×10^{8}	meters
L_S = Sun luminosity:	3.90×10^{26}	watts
T_S = Sun effective surface temperature:	5778	Kelvin

Source: E. Chaisson and S. McMillan, *Astronomy Today*, 6th edn. (Pearson Addison-Wesley, San Francisco, CA, 2008).

APPENDIX 5: SOME PHYSICS QUANTITIES AND EQUATIONS

This appendix is designed to be a refresher for those who have been away from the sciences for a while, but would like to follow or check the mathematical assumptions in text.

Scalars and vectors

Scalar quantities, such as mass, have magnitude (or size only). Vectors (such as velocity) have size and direction.

Displacement, velocity, and acceleration

Displacement is the distance traveled, often represented by x. Units in the MKS system are meters. Velocity is the change of displacement with time, often represented by $v = \Delta x/\Delta t$, where Δ is the Greek upper case delta and represents "change in". The unit of velocity in the MKS system is meters/second. Acceleration is the change of velocity with time, often represented by $a = \Delta v/\Delta t$. The MKS unit of acceleration is meters/second^2. Examples of acceleration include gravitational acceleration at Earth s surface, $g = 9.8\ \text{m/s}^2$, and centripetal acceleration (the acceleration of an object constrained to move in a circle), a_{cent}.

Force, mass, and weight

The force on an object F is an action that alters the motion of an object. Newton defined force as $F = ma$, where m is the object's mass and a is the object's acceleration due to the force. In the MKS system, the unit of force is the Newton. Mass is an intrinsic property of matter. It can be thought of as the amount of material in an object or the resistance of the object to changes in its motion (inertia). The MKS unit of mass is kilograms. Weight is the force of Earth's gravitational attraction on an object. An object's weight W_{obj} can be written as $W_{\text{obj}} = M_{\text{obj}}g$.

Gravitational force

The gravitational force between two objects is always attractive. Consider two objects A and B. The center-to-center distance between the two objects is R and their masses are M_A and M_B. The mutual gravitational force between A and B is $F_{\text{grav}} = GM_AM_B/R^2$, where G is the gravitational constant. If A is an object on Earth's surface and B is the Earth, the gravitational acceleration at Earth's surface can be written as $g = GM_E/R_E^2$, where R_E is Earth's radius and M_E is Earth's mass.

Centripetal force

The force necessary to keep an object with mass M moving in a circular path with radius R and velocity V is the centripetal force. Its magnitude is calculated from $F_{\text{cent}} = MV^2/R$.

Work and energy

The work done by a force F on an object as it moves through a distance ΔX is defined as $W = F\,\Delta X$. The work done on an object changes the object's kinetic energy, KE, which is defined $KE = \frac{1}{2}MV^2$, where M is the object's mass and V is the object's velocity. If an object is at a height h above a reference frame near Earth's surface, the potential energy PE can be expressed by $PE = mgh$. Even if an object has no kinetic energy and no potential energy, it still has rest energy E_R. This is defined using Einstein's famous equation: $E_R = mc^2$, where m is the object's mass and c is the speed of light in vacuum. The total energy of any object—the sum of kinetic, potential, and rest energy—is conserved. Because rest energy is derived from the conversion of matter into energy, this conservation principle is often referred to as "the conservation of mass energy." Work and energy are measured in Joules in the MKS system.

Power

The rate at which the energy of a system is expended with time is called the power. Power is the ratio of energy/time. The MKS unit is Joules/second. One Joule per second is defined as 1 watt.

Linear momentum

Another conservation law deals with the total linear momentum of a system. The linear momentum of an object is defined as the product of that object's mass and velocity. Linear momentum is a vector quantity. In any interaction between or among the objects in a system (such as a collision), the individual linear momentum of the particles can be redistributed. However, the total linear momentum of the system is conserved. It is interesting to note that a special MKS unit has never been identified for linear momentum. In the MKS system, therefore, the value of an object's linear momentum is given in units of kilograms-meters/second.

Source: Any college or university level physics text could be used to compile the information supplied in this appendix. I have used: H. C. Ohanian, *Physics*, 2nd edn. (Norton, NY, 1989).

APPENDIX 6: ESTIMATING THE MASS OF THE INNER GALAXY

This appendix makes use of some of the information presented in Appendices 1–5 and applies it to the problem of estimating the mass and number of stars in the inner portion of the Milky Way: that part of our galaxy closer to the center than the orbit of the Sun.

Solar galactic orbit

According to Alan Fiala, the Sun's distance from the center of the Milky Way galaxy is about 25,000 light years. Since 1 light year is about 10^{16} meters, the Sun's approximate distance from the galactic center in MKS units is $R_{gc} \approx 2.5 \times 10^{20}$ meters. The Sun's velocity in the direction of its revolution around the galactic center is about 237 kilometers per second. In the MKS system, this is expressed as $V_{\mathrm{Sun}} \approx 2.37 \times 10^5$ meters per second.

Equating gravitational and centripetal forces

Assuming that the Sun's galactic orbit is perfectly circular, the gravitational force between the galaxy closer to the galactic center and the Sun is equated to the centripetal force on the Sun. Substituting and manipulating, we obtain the following expression for the mass of the galaxy closer to the galactic center than the Sun:

$$M_{IG} = \frac{V_{\mathrm{Sun}}^2 R_{gc}}{G},$$

where G is the gravitational constant (6.67×10^{-11} Newton-meter^2/kilogram^2).

Substituting in this result for the Sun's velocity and distance from the galactic center, we calculate: $M_{IG} \approx 2.11 \times 10^{41}$ kilograms. However, the mass of the Sun is about 2×10^{30} kilograms. So, the mass of the inner galaxy is approximately equal to that of 10^{11} or one hundred billion Suns.

Estimating the number of stars in our galaxy

Since the Sun is about halfway between the galactic center and the galactic hub, the total mass of the Milky Way galaxy is about equal to 200 billion Suns. However, the Sun is more massive than about three quarters of the stars. So an estimate of 350 billion stars in the Milky Way galaxy is not too far off.

Source: A. D. Fiala, "General astronomical constants," chap. 2 in *Allen's Astrophysical Quantities*, 4th edn. (ed. A. N. Cox, Springer-Verlag, NY, 2000).

Index